Cross-Layer Approximation and In-Network Acceleration

Zahra Ebrahimi · Akash Kumar

Cross-Layer Approximation and In-Network Acceleration

Enabling the Next Generation of Sustainable and High-Performance Reconfigurable Systems

Springer

Zahra Ebrahimi
Faculty of Electrical Engineering
and Information Technology
Ruhr University Bochum
Bochum, Germany

Akash Kumar
Faculty of Electrical Engineering
and Information Technology
Ruhr University Bochum
Bochum, Germany

ISBN 978-3-032-21711-0 ISBN 978-3-032-21712-7 (eBook)
https://doi.org/10.1007/978-3-032-21712-7

This Springer imprint is published by the registered company Springer Nature Switzerland AG
The registered company address is: Gewerbestrasse 11, 6330 Cham, Switzerland

Dedicated to our families, and the ones who tirelessly strive for the sustainability of our beautiful and unique planet.

Preface

Over the past decades, energy efficiency and sustainability—together with high performance and throughput—have continued to be central design priorities in the computing industry. Traditionally, these goals were achieved by exploiting technology scaling (in line with Moore's law) and by applying microarchitectural optimizations. However, with the breakdown of Dennard scaling, the energy and performance benefits derived from these methods have steadily diminished. As a result, new design strategies have become essential to satisfy the computational and energy demands of modern applications, each with its own unique characteristics. In many signal processing workloads and contemporary applications running across the edge-to-cloud spectrum—from IoT devices to largescale data centers—three properties have become increasingly apparent. First, these workloads operate on real-world data that is frequently noisy or repetitive. Second, their intermediate computations tend to tolerate inaccuracies. Third, instead of producing one exact "golden" output, these applications often yield multiple acceptable solutions.

To exploit such inherent error tolerance, the paradigm of Approximate Computing has gained prominence. This approach relaxes strict correctness requirements to achieve high performance and energy-efficient computation while maintaining acceptable output quality. In particular, Cross-Layer Approximation capitalizes on the combined benefits of techniques applied across multiple abstraction levels (e.g., circuit, architecture, and application layers) to maximize performance improvements.

Historically, most approximation techniques have been developed for Application-Specific Integrated Circuits (ASICs). Yet, ASICs suffer from long development cycles and substantial Non-Recurring Engineering (NRE) costs. Reconfigurable platforms such as Field-Programmable Gate Arrays (FPGAs) and Coarse-Grained Reconfigurable Architectures (CGRAs) mitigate these drawbacks. Nevertheless, direct adoption of ASIC-oriented approximation methods on reconfigurable hardware fails to deliver proportional performance gains due to fundamental architectural differences. Therefore, approximation strategies must be explicitly designed or adapted for reconfigurable fabrics. Moreover, most circuit-level approximation work has concentrated on addition and multiplication. Although less frequent, division remains unavoidable in many error-resilient workloads. Division operations incur higher area and energy costs and exhibit significantly greater latency than multiplication, often limiting overall application performance. These challenges also make divider integration within CGRA Processing Elements (PEs) problematic. Incorporating a resource-intensive divider into a PE substantially increases area, power consumption, and reduces operating frequency. Consequently, many approximate CGRA architectures omit divider support entirely, preventing division-containing kernels from being offloaded to the CGRA and forcing the host processor to remain involved for these operations. Furthermore, nearly all existing approximate multipliers and dividers are SISD and nonpipelined, leaving the potential for higher throughput untapped. In addition, many state-of-the-art cross-layer approximation approaches neglect architectural-level optimization opportunities and fail to construct a cohesive stack of techniques across abstraction layers. Existing cross-layer approaches for multi-kernel applications have primarily targeted Neural Networks (NNs), leaving numerous other domains unexplored. A general methodology capable of jointly tuning multiple approximation knobs across layers for multi-kernel workloads is still missing. A key prerequisite—kernel-level error-resilience sensitivity analysis quantifying end-to-end performance versus Quality of Results (QoR) trade-offs—has also been largely overlooked.

The central objective of this work is to address these challenges by developing high-throughput, energy-efficient approximate architectures and techniques using a cross-layer approach tailored to reconfigurable fabrics. To achieve this, our methodology leverages synergistic techniques across both the data acquisition phase and the computation phase of multi-kernel applications. These techniques operate throughout the computing stack, spanning circuitlevel, architectural-level, and application-level abstractions. Our methodology proceeds in two major steps. First, kernel-wise sensitivity analysis quantifies how individual kernels respond to different approximation techniques, capturing the relationship between achievable end-to-end performance gains and the associated QoR degradation. Second, a greedy heuristic adjusts approximation levels (i.e., cross-layer knobs) to maximize performance while satisfying user-defined QoR constraints. We further extend this methodology to support energy-efficient cross-layer approximation for distributed and in-network computing (INC) within emerging 5G/6G systems.

We introduce several approximate designs for multiplication and division—the two most critical and computationally intensive arithmetic operations in modern workloads such as signal, image, audio, and bio-signal processing, as well as Machine Learning (ML). To enhance throughput and energy efficiency, our designs incorporate fine-grained pipelining and Single Instruction, Multiple Data (SIMD) execution. We propose a hybrid SIMD multiplier–divider module that supports both operations at runtime without reconfiguration. This design synergistically combines precision scaling with approximate multiplication and division, while taking advantage of SIMD at the architectural level—forming a coherent cross-layer approximation framework. The proposed SIMD multiplier–divider is further integrated into PE architectures to support SIMD/MIMD execution in CGRAs. The operating mode of each PE (SISD, SIMD, or MIMD) is selected according to our cross-layer approximation methodology. The precision-tuning heuristic determines the bit-width needed for each kernel mapped onto one or more PEs, based on sensitivity analysis results, ensuring acceptable QoR. Finally, we extend our cross-layer methodology to enable in-network acceleration of multi-kernel applications within chains of resource-constrained 5G/6G network nodes. In this setting, we first restructure complex application graphs to approximate unsupported arithmetic operations within programmable switches. By incorporating additional application-level techniques and reducing sensing frequency or sampling periods, our methodology decreases resource consumption, lowers communication overhead, and supports real-time responsiveness.

Bochum, Germany

Zahra Ebrahimi
Akash Kumar

Contents

Acronyms

1D	1-Dimensional
2D	2-Dimensional
AA	Affine Arithmetic
ADC	Analog-to-Digital Converter
ADP	Area-Delay Product
AI	Artificial Intelligence
ALM	Adaptive Logic Module
ALU	Arithmetic Logic Unit
ARE	Average of absolute Relative Error
ASAM	Adaptive Statistical Approximation Model
ASIC	Application Specific Integrated Circuit
BB	Basic Block
BSS	Blind Source Separation
CGRA	Coarse-Grained Reconfigurable Architecture
CLA	Carry Look Ahead
CLB	Configurable Logic Block
CNN	Convolutional Nural Network
CPU	Central Processing Unit
DCT	Discrete Cosine Transform
DFG	Data Flow Graph
DLP	Data Level Parallelism
DMA	Direct, Memory Access
DNN	Deep Nural Network
DoA	Degree of Accuracy
DoF	Degree of Freedom
DPR	Dynamic Partial Reconfiguration
DPU	Dataflow Processing Unit
DSE	Design Space Exploration
DSP	Digital Signal Processing
DT	Decision Tree
DVFS	Dynamic Voltage and Frequency Scaling

DWT	Discrete Wavelet Transform
EB	Error Bias
ECG	Electrocardiogram
EEG	Electroencephalogram
FA	Full Adder
FF	Flip Flop
FFT	Fast Fourier Transform
FIR	Finite Impulse Response
FP	Floating-Point
FP64	Double Precision Floating-Point
FPGA	Field Programmable Gate Array
FSM	Finite-State Machine
GA	Genetic Algorithm
GPP	General Purpose Processor
GPU	Graphics Processing Unit
HA	Half Adder
HCD	Harris Corner Detection
HDL	Hardware Description Language
HEVC	High Efficiency Video Coding
HLS	High-Level Synthesis
HPF	High-Pass Filter
IA	Interval Arithmetic
IC	Integrated Circuit
ICA	Independent Component Analysis
IIR	Infinite Impulse Response
ILP	Integer Linear Programming; Instruction Level Parallelism
IMT	International Mobile Telecommunications
INC	In-Network Computing
IoT	Internet of Things
IP	Intellectual Property
IPC	Instruction Per Cycle
IPU	Infrastructure Processing Unit
ISA	Instruction Set Architecture
JPEG	Joint Photographic Experts Group
KNN	K-Nearest Neighbor
KPI	Key Performance Indicator
LLM	Large Language Model
LLVM	Low Level Virtual Machine
LOD	Leading One Detector
LSB	Least Significant Bit
LUT	Look-up Table
MAC	Multiply-ACcumulate
MEC	Multi-Access Computing Error
MIMD	Multiple Instruction, Multiple Data
MIMII	Malfunctioning Industrial Machine Investigation and Inspection

ML	Machine Learning
MLP	Multi-Layer Perception
MNIST	Modified National Institute of Standards and Technology
MRED	Mean of Absolute Relative Error
MSB	Most Significant Bit
MSE	Mean Square Error
MTTR	Mean Time To Respond
NIC	Network Inference Card
NN	Neural Network
NPU	Network Processing Unit
NRE	Non-Recurring Engineering
PE	Processing Element
PMF	Probability Mass Function
PP	Partial Product
PRE	Peak of absolute Relative Error
PSNR	Peak Signal-to-Noise Ratio
QoR	Quality of Result
RAM	Random Access Memory
RF	Random Forest
RISC-V	Reduced Instruction Set Computer-V
RMS	Recognition, Mining, and Synthesis
RMSE	Root Mean Square Error
RoI	Region of Interest
ROM	Read Only Memory
RTL	Register Transfer Level
SAM	Statistical Approximation Model
SC	Stochastic Computing
SDN	Software-Defined Networking
SDR	Signal-to-Distortion Ratio
SEU	Single-Event Upset
SFU	Special Function Unit
SIMD	Single Instruction, Multiple Data
SISD	Single Instruction, Single Data
SNR	Signal-to-Noise Ratio
SoC	system on Chip
SRAM	Static Random, Memory Access
SSIM	Structural Similarity Index
SVM	Support Vector Machine
UAV	Uunmanned Aerial Vehicle
WLO	Word-Length Optimization
XPE	Xilinx Power Estimator

Chapter 1
Introduction

Abstract Over recent decades, growing algorithmic complexity has outpaced the benefits of Moore's Law and Dennard scaling, as power density, leakage, and thermal limits prevent further gains from frequency increases and simple transistor scaling. Multi-core processors and Dark Silicon techniques mitigate power constraints, but parallelism limits and memory bottlenecks restrict achievable speedups, especially under strict energy and real-time requirements. In parallel, many modern workloads are inherently error-resilient, accepting multiple satisfactory outputs instead of a single exact result, which opens the door to approximate computing. By relaxing accuracy where tolerable, approximation can significantly reduce computation, memory, and energy costs while improving performance, and has shown order-of-magnitude gains in both industrial and academic studies. The text positions approximate computing as a key enabler of sustainable, energy-efficient systems from edge devices to cloud servers. It then highlights a gap: most early techniques target ASICs and transfer poorly to reconfigurable fabrics such as FPGAs and Coarse-Grained Reconfigurable Architectures (CGRAs). The book addresses this by proposing cross-layer, sensitivity-guided approximation, and distribution methods tailored to reconfigurable and in-network computing platforms, particularly for multi-kernel and 5G/6G scenarios.

Keywords Cross-layer approximation · In-network computing · Distributed computing · 5G-6G · Embedded systems · FPGA · CGRA · SIMD/MIMD · High-throughput · Pipeline · Sustainability · Energy-efficiency · Error sensitivity analysis · Multi-objective heuristic · Design space exploration · Programmable networks

1.1 Introduction

Over the past few decades, the computational demands required to implement a broad range of applications have grown exponentially, driven by the steadily increasing complexity of emerging algorithms. In response, the semiconductor industry

Z. Ebrahimi and A. Kumar, *Cross-Layer Approximation and In-Network Acceleration*,
https://doi.org/10.1007/978-3-032-21712-7_1

has continually attempted to address this rising computational pressure by producing successive generations of Integrated Circuits (ICs) with improved processing capabilities. These gains in computing speed are typically associated with Moore's Law, which states that advances in technology scaling enable nearly twice as many transistors to be fabricated on a chip of the same size approximately every two years [Moo98]. Although higher transistor density can enhance performance, actual speed improvements depend on multiple factors, such as architectural enhancements, power efficiency, and interconnect scaling. Nevertheless, maintaining the transistor-doubling trend predicted by Moore's Law has become increasingly challenging and expensive, particularly at technology nodes such as 5 nm [Boh09].

In parallel, Dennard scaling asserts that as transistors shrink, their power density remains constant because operating voltage and current decrease proportionally with device size. This effect would theoretically allow more transistors to be integrated within the same chip area without increasing total power consumption [Den+74]. However, growing leakage currents, rising power density, and thermal instability have led to the breakdown of Dennard scaling in 65 nm and more advanced technologies [Mar14]. Consequently, relying solely on higher operating frequencies or resource over-provisioning is no longer sufficient for designers seeking improved performance per watt.

As a result, the industry shifted toward multi-core processor designs as the next means of achieving high-performance computing. Yet even with multi-core architectures, it is not feasible to operate all cores at maximum power simultaneously. The practice of selectively power-gating idle cores to reduce total chip power is known as Dark Silicon [Esm+11]. Moreover, despite using multiple cores, many applications yield limited speedup due to parallelism challenges such as data dependencies, serialization, synchronization overheads, and memory bandwidth limitations [Ven+15b]. Meeting rapidly increasing computational requirements—and enabling real-time responsiveness, especially on battery-powered devices with limited processing capability—therefore demands innovative *computing* approaches to further enhance system performance.

It is also important to note that, alongside technological and architectural developments, the nature of computing workloads has evolved considerably. A defining feature of many modern applications is that they can produce several acceptable outputs rather than a single deterministic result. For example, a search engine may return multiple suitable responses instead of one "best" answer. Likewise, audio and video processing applications can use different encoders to generate a range of acceptable outputs for mobile devices. Applications that yield multiple feasible solutions, rather than a single exact one, inherently exhibit resilience to errors in their intermediate computations.

This inherent tolerance to inaccuracy has led to the rise of *approximate computing* as a promising strategy for building fast and energy-efficient systems [SG17]. By intentionally introducing controlled approximations, this paradigm trades strict accuracy for performance-related benefits such as reduced computational and memory demands, lower energy use, and faster execution [Mit16, Sta+20]. The importance of approximation has grown substantially in recent years across both academia and industry [LLS20]. Indeed, approximation has become central to addressing the escalating need for resource efficiency and real-time processing across systems spanning the edge-to-cloud continuum—from battery-powered embedded devices to high-performance servers—where energy efficiency and responsiveness are crucial. Prominent industrial (e.g., Intel, IBM, Microsoft, and Bosch [Kar+19, Ven+20, Ven+21b, SGA19, Wan+18]) and academic [PRS19a, SKi+19, HKS19, Jia+18] efforts have shown that approximation can achieve more than an order-of-magnitude performance improvement, a level of gain rarely attainable through traditional techniques such as multi-core processing or low-power design.

Beyond performance benefits, approximate computing is also aligned with the principles of sustainable computing. By reducing the energy consumption of computation-intensive tasks, approximation helps build environmentally conscious systems that conserve resources and lower carbon footprints. This aligns directly with the goals of sustainable computing, which emphasize energy efficiency and minimal ecological impact. Furthermore, sustainable computing promotes reusability and reconfigurability of both software and hardware. Integrating approximate computing with reconfigurable systems therefore offers a promising pathway to sustainable solutions by meeting increasing computational needs without significantly escalating energy consumption or environmental cost.

Early research on approximate computing primarily focused on developing and evaluating techniques—targeting predominantly circuit-level or application-level optimizations—for Application Specific Integrated Circuit (ASIC)-based platforms. As a result, these techniques did not deliver comparable gains when directly deployed on reconfigurable platforms such as Field Programmable Gate Arrays (FPGAs) and CGRAs . This mismatch arises from the fundamental architectural differences between ASICs and FPGAs [Ull+18, SSA18]. The present book, together with other parallel works from the last six years [SUl+20, Ull+18, Ull+21, Pra+18, Ull+22], belongs to the early set of studies that recognized this limitation and aimed to design approximation techniques specifically tailored to reconfigurable systems.

This book differentiates itself from prior work by focusing on designing cross-layer techniques that target high throughput and energy efficiency on reconfigurable fabrics. To this end, it proposes a set of approximation and optimization strategies and imprecise architectures tailored to both FPGA- and CGRA-based platforms. These techniques operate across the data acquisition and computation phases of multi-kernel applications and span various abstraction layers—ranging from the circuit level to architectural and application levels. A sensitivity analysis method is introduced to quantify the error resilience of each application kernel to different approximation strategies. Based on this analysis, a greedy heuristic is developed to efficiently tune cross-layer approximation knobs. The methodology is then extended to 5G/6G networks, particularly targeting distributed or INC environments. The proposed cross-layer approximation and distribution framework is evaluated using an industrial application case study, demonstrating that it can successfully execute multi-kernel workloads on resource-constrained network nodes—either centrally or in distributed fashion—while enabling real-time responsiveness.

The remainder of this chapter is organized as follows. Section 1.2 defines the inherent error resilience of applications and outlines the factors contributing to this resilience. Section 1.3 introduces the approximate computing paradigm for error-resilient workloads and summarizes key state-of-the-art contributions across the computing stack. Section 1.5 discusses the importance of reconfigurable fabrics and reviews techniques aimed at improving energy efficiency and achieving high throughput on FPGA and CGRA platforms. Since higher throughput and energy efficiency depend on understanding the cost of accurate arithmetic, Sect. 2.3 details performance metrics for widely used or complex operations such as addition, multiplication, division, and square root . Section 1.6 then introduces the concept of In-Network Computing (INC) and explains how approximation can enable multi-kernel applications to run on resource-limited network nodes, enabling real-time acceleration during in-network data processing. Section 1.7 elaborates on the research challenges addressed in this book. Section 1.8 summarizes the key contributions. Finally, Sect. 1.9 provides an overview of the book's structure and the connections between its chapters.

1.2 Inherent Error Resilience of Applications

This concept refers to an application's ability to produce outputs of acceptable quality despite some degree of inaccuracy in its intermediate computations or final results. The term *acceptable quality* means that different quality levels may still satisfy user or system requirements in various deployment scenarios, rather than requiring a single precise output. For instance, metrics such as Peak Signal-to-Noise Ratio (PSNR) and Signal-to-Distortion Ratio (SDR) are widely used to assess the quality of image and audio processing , respectively. Consequently, the appropriate accuracy thresholds vary across use cases—from mobile devices with strict resource constraints to high-performance cloud platforms with more relaxed resource budgets.

Error-resilient applications appear across numerous domains, including signal processing (e.g., image/video, audio, and bio-signal), Machine Learning (ML) models, biometric authentication (e.g., iris scanning), big data analytics, networking and communication protocols, and Recognition, Mining, and Synthesis (RMS) applications. Works identifying error-resilient portions of such applications are surveyed in Sect. 3.7 of Chap. 3. Prior research shows that these applications spend, on average, 70–83% of their energy on computations that are inherently error-tolerant [Yaz+17]. The sources of error resilience in applications stem from several factors [Chi+13b], which can be grouped into three categories:

- **Inputs**: Real-world data are often noisy and contain redundancy. For example, audio virtual assistants such as Siri, Google Assistant, and Alexa must process acoustic signals in environments with overlapping noise sources. Consequently, algorithms designed for such data are inherently robust to noise and input variations.

- **Computations**: Some applications—such as ML models—rely on probabilistic computations and are therefore naturally error-resilient. In addition, many algorithms employ iterative or training-based methods to refine their initial estimates, progressively reducing error in subsequent iterations. Several classes of applications (e.g., low-bit-width Neural Network (NN) training [Zho+18, CBD15] or audio anomaly detection [Wu+21d, Wu+21c, Wu+22a, Wu+21a]) deliberately leverage this property to absorb or manifest errors during execution.
- **Outputs**: Many applications produce multiple feasible outputs rather than a single golden solution. For instance, an image may be correctly labeled as a "cat," a "mammal," or more broadly, an "animal." Likewise, search engines typically return multiple possible matches for a user query. Moreover, human perceptual limitations—e.g., insensitivity to minor noise or compression artifacts in images or audio—further widen the acceptable range of output Quality of Result (QoR).

It is crucial to note that not all applications are error-resilient. For example, an improper use of precision scaling from 32-bit to 16-bit (an approximation technique discussed later) caused failures that contributed to the destruction of the Ariane 5 rocket in 1996 and issues in the Boeing 787 aircraft in 2013 [BBC15]. Furthermore, not all components within an error-resilient application are themselves resilient to approximation. Control-flow logic, memory index calculations, and loop counters often require the highest accuracy and precision; otherwise, errors can escalate dramatically. Such components must therefore be executed in accurate mode even when other parts of the application can tolerate approximation.

1.3 Approximate Computing Paradigm and Techniques

The goal of approximate computing is to trade computational accuracy for improvements in performance, such as increased speed, higher throughput, or reductions in an application's resource footprint. Rather than always targeting the highest possible accuracy—which can be costly and delay responses—approximate computing aims to deliver results that fall within an acceptable quality range. Before exploring this concept further, it is useful to distinguish approximate computing from stochastic computing and fault-tolerant (error-resilient) computing, which are discussed below.

Fault-Tolerant Computing: Ensuring the reliability of ICs remains a major challenge in advanced semiconductor technologies. As technology scaling continues and transistors become smaller, they become increasingly susceptible to various types of faults and soft errors. These faults—either transient or permanent—can originate from numerous sources, including hardware issues (e.g., device degradation or corrupted memory cells) or software problems (e.g., logic bugs, race conditions). Transient faults typically arise from environmental disturbances such as electrical noise or voltage fluctuations and may appear as Single-Event Upsets (SEUs) or other soft errors. Permanent faults, by contrast, stem from physical defects or wear-out in hardware components, such as faulty processor cores or damaged memory arrays.

To detect, isolate, and recover from such faults, a wide range of reliability-enhancing techniques have been proposed under the umbrella of fault-tolerant computing. Some techniques reduce operating frequency, increase supply voltage, use redundant execution units, or incorporate sophisticated error-correction mechanisms [DBW15]. However, these solutions often degrade system performance in exchange for reliability. More recent efforts have also investigated architectural- and application-level masking mechanisms to improve system performance while still maintaining acceptable reliability [Lee+10].

Stochastic Computing (SC): Approximate computing and SC both aim to improve system performance, yet they fundamentally differ in principles, goals, origin, and error behavior. *Principle*: Unlike traditional binary computing, which uses deterministic bit patterns, SC relies on random binary streams in which the value is encoded by the proportion of 1s in the bitstream. Thus, a single number may have multiple valid stochastic representations [NPH19, VA19]. For instance, the value 0.5 may be expressed as 0101010100 or 01101001 in an 8-bit format. Each bit therefore reflects a probability of being '1' or '0'. *Source of error*: In contrast to approximate computing, where errors are intentionally introduced, errors in SC arise from randomness. Since each bit represents a probability rather than a fixed value, statistical fluctuations cause nondeterministic behavior. This makes SC inherently probabilistic, whereas all approximation techniques discussed in this book operate deterministically. *Error handling*: Random variations in SC are managed using statistical techniques such as averaging across multiple samples or applying error-correcting methods. *Goal*: Due to its probabilistic nature, SC naturally exhibits strong fault tolerance. Combined with the fact that bit position is often irrelevant, SC is especially useful in applications where robustness to noise or faults is essential.

Approximation Techniques Across Application Phases: Techniques developed under approximate computing exploit application-level error resilience to trade accuracy for performance gains. Many such techniques—explicitly or implicitly—have been employed since the early days of digital computing. For example, Analog-to-Digital Converters (ADCs) represent one of the earliest forms of approximation: they convert continuous real-world signals into discrete digital values of user-selected bit-width. Such approximate data conversion has produced a variety of numeric representations, including fixed-point and Floating-Point (FP) formats, both of which have been used for decades [Wik24].

According to the National Science Foundation [US 25], application execution in cyber-physical and Internet of Things (IoT) systems is broadly divided into three phases: *data acquisition* (physical layer), *computation* (cyber layer), and *communication* (network layer) [WZW21]. This book focuses primarily on the first two phases, employing approximation techniques across both data acquisition and computation[1].

[1] Approximation can also be applied at the communication layer using techniques such as dynamic bitrate reduction or probabilistic packet dropping.

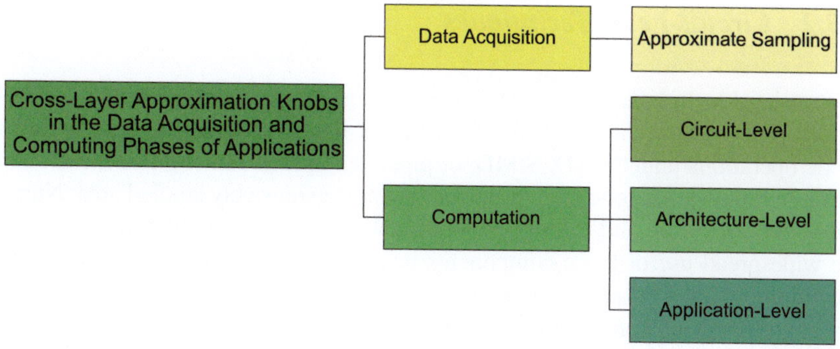

Fig. 1.1 Classification of approximation techniques in the data acquisition (sensing) and computing (processing) phases of applications

Figure 1.1 provides an overview of widely used approximation techniques for the data acquisition and computing phases.[2] As illustrated, inaccuracies can be introduced at multiple levels—ranging from circuit-level approximations to architecture- and application-level optimizations. Some works also explore combining approximation across multiple layers, known as *cross-layer* approximation [MSh+16]. The following subsections briefly describe the most commonly applied techniques; a more detailed discussion appears in Chap. 3.

1.3.1 Approximate Data Acquisition (Sampling) Techniques

Sampling strategies are generally categorized into random and non-random methods [Dal+24]. While random sampling selects data points purely based on probability without any imposed structure, non-random approaches introduce organization or constraints into the selection process to ensure that key properties of the signal are preserved. Non-random techniques are further classified as *stratified* or *adaptive*. Stratified sampling partitions the signal or dataset into subgroups (strata) and draws samples from each subgroup proportionally to maintain representative coverage. Adaptive sampling, in contrast, adjusts its selection dynamically based on data characteristics, often prioritizing more informative regions. As will be further discussed in Chap. 3, this book adopts a stratified sampling technique to preserve the reconstructability of the original signal.

[2] A more extensive classification is presented in Chap. 3.

1.3.2 Circuit-Level Techniques

Circuit-level techniques include Dynamic Voltage and Frequency Scaling (DVFS), circuit pruning, and the deployment of inexact hardware configurations for arithmetic operations under SISD, SIMD, or pipelined modes. Among these, the design of approximate arithmetic units has been the most extensively studied area. Numerous imprecise adder and multiplier configurations exist in the literature, owing to the widespread use of these components in error-resilient applications. For example, multiplication is among the most resource-intensive operations, accounting for nearly 99% of computational energy in NNs [Jai+18]. A single 224 × 224 input image processed by ResNet152 or VGG-16 requires 11.3 and 15.5 billion Multiply-ACcumulate (MAC) operations, respectively [Sze+17, He+16].

While the majority of circuit-level approximation efforts have focused on multipliers, studies such as [EUK20, HHS19] and our own analysis (Sect. 2.2) show that division has significantly higher latency and energy cost than multiplication, making it a performance bottleneck. Consequently, interest in approximating division has grown, since division—although less frequent—is unavoidable in many error-resilient computations.[3] Below, key hardware-based approaches for designing inexact arithmetic units are outlined. A more detailed discussion appears in Sect. 3.3.

Adders: Three main approaches are used to design approximate adders [Bec+16, Pra+18, Ull+18, SSA18]. First, for performance, the long carry chain of an N-bit adder is segmented into k-bit sub-adders that operate in parallel. Second, for area efficiency, the Karnaugh maps of Full Adders (FAs)/Half Adders (HAs) are simplified for Least Significant Bit (LSB) addition in higher-order adders. Third, configurable-accuracy adders integrate inexact sub-adders with error detection and correction units. In this book, we use accurate adders throughout, based on three observations. First, as detailed in Chap. 2 (Sect. 2.2), multiplication, division, and square root operations consume significantly more area, latency, and energy (2–1700×) than adders of the same width. Second, the cumulative energy of addition is relatively small compared to other Arithmetic Logic Unit (ALU) operations across the studied signal-processing workloads. Third, studies such as [HJi+20] show that approximating additions leads to larger accuracy degradation than approximating multiplications in matrix-multiplication-heavy tasks such as image/video, audio, and NN processing.

Multipliers: Four general methodologies are used for designing approximate multipliers. (1) Approximation applied to the generation, addition, or accumulation of Partial Products (PPs), e.g., carry truncation in adders or compressors [YHS20, Ull+18, VL20, MSt+20, JSD20, DEs+18, MWa+20, SAn+18]. (2) Resizing a high-order multiplier into a smaller unit by performing Most Significant

[3] For example, K-means clustering, Discrete Cosine Transform (DCT) in JPEG, and AlexNet Convolutional Nural Network (CNN) require division extensively [Kri+12].

Bit (MSB)-only multiplication [SVa+19, RZe+17, HJi+19, HJi+18]. (3) Designing small, low-gate-count imprecise multipliers (e.g., 2×2 units using simplified Karnaugh maps) and assembling them modularly [SSA18, VMr+17, SRe+16]. (4) Using Mitchell's algorithm [NMi62], which transforms multiplication (and division) into addition (or subtraction) in the logarithmic domain, replacing the 2-Dimensional (2D) PP array with shifts and a 1-Dimensional (1D) adder. This often yields higher performance, though the method is inherently approximate. Several works propose error-reduction enhancements for Mitchell-based designs [SFJ19, WLi+18, HHS18a, HSa+20, HHS19, ZSA20]. Mitchell's algorithms form the basis of the imprecise units introduced in this book and are further explained in Chap. 2 (Sect. 2.4), with related works discussed in Sect. 3.3.

Dividers: Although less frequent, division remains essential in most error-resilient workloads and is often the slowest and most energy-intensive ALU operation. As a result, approximate division has received increased attention. Existing approaches fall into four main categories.

(1) Computing the division using only the MSBs of operands [HJi+19, HJi+18], though errors grow when too many bits are truncated.
(2) Employing imprecise subtractors in classical shift-and-subtract division [ESS20, SES19, LCh+18, LCh+16], but high-order inexact subtraction leads to large errors.
(3) Computing an approximate reciprocal of the divisor and multiplying it by the (possibly truncated) dividend [Vae+18, SVa+17, RZe+16, Mel+19], though reciprocal computation (e.g., via Taylor or Newton-Raphson) is iterative and costly.
(4) Using Mitchell's logarithmic division [HHS19], which generally yields higher performance than traditional approaches.

1.3.3 Architecture-Level Techniques

Inexact Computing Architectures: One line of research proposes approximate configurations for Digital Signal Processing (DSP) units, ALUs, or Processing Elements (PEs), which are then used to build larger systems such as FPGAs or CGRAs. Several CGRA works [Bra+18, Akb+18, Akb+20] have replaced accurate adders, multipliers, and dividers within PEs with one or more approximate variants, selected dynamically at runtime. However, this intra- or inter-PE heterogeneity introduces additional logic and routing overheads. Other studies have modified FPGA DSP units to operate as dual multipliers with shared operands [LGB20, Fu+17a]. Quora [Ven+13], a quality-programmable vector processor, supports three accuracy levels via a specialized Instruction Set Architecture (ISA) and employs arrays of approximate and mixed-accuracy PEs alongside one accurate PE. Further architectural-level techniques are discussed in Sect. 3.4.

Inexact Storage (Cache) Architectures: Another research track focuses on approximate storage mechanisms for caches. Even with multi-level cache hierarchies, accessing main memory incurs high latency and energy costs, and cache misses are common. To mitigate this, prior works propose techniques ranging from approximate read/write operations [Ran+17] to generating approximate values on cache misses [Yaz+16, Mig+16]. Providing an approximate value enables continued execution without stalling on memory access. These works are discussed further in Chap. 3.

1.3.4 Application-Level Techniques

This category includes precision scaling, data reuse and memoization, computation skipping and task perforation, and modifications of application control flow or programming models. Among these, precision scaling, computation skipping, and task perforation have been the most widely explored; the remaining techniques are discussed in Sect. 3.5.

Precision Scaling: Historically, computing systems—from embedded devices to high-performance servers—have performed 32- or 64-bit integer or FP operations. However, many signal-processing and machine-learning workloads do not require such high precision. Large data widths increase computational cost, memory footprint, and implementation difficulty on embedded platforms. Precision scaling, therefore, has become a major approximation technique, with extensive studies demonstrating the viability of 16- and 8-bit precision, and recent research exploring sub-8-bit quantization for NNs [Yin+17, CBD15, Moo+17, Vog+19, LTA16]. In signal-processing tasks, precision reduction may be applied to both input data and kernels or filter coefficients.

Computation Skipping and Task Perforation: These techniques prove effective for memory- or computation-heavy workloads. Loop perforation—skipping loop iterations—is widely used in general-purpose settings. Increasing convolution filter stride is another approach that inherently skips computation. In NNs, neuron pruning (at neuron or filter level) and early classification are prevalent examples of computation skipping.

1.3.5 Cross-Layer Techniques

Approximate computing principles can also be applied across multiple layers of the computing stack, known as cross-layer approximation [MSh+16, Chi+14]. Combining approximation at different layers often yields significantly higher performance than applying a single technique in isolation. The aim of cross-layer strategies is to exploit approximation opportunities consistently across application-, architecture-, and circuit-level abstractions within a unified hierarchy. Most prior work has focused on cross-layer approximation for NNs, with fewer studies exploring signal-processing areas such as image or bio-signal processing [Ebr+21, USK21, Ven+20, Gon+19, SSa+18, Ven+15a, Fan+19b, Han+18]. Neural network-oriented methods mainly combine approximate multipliers (circuit level) with quantization and pruning (application level). Relevant studies are summarized in Sect. 3.6.

A complete cross-layer methodology requires error-sensitivity analysis, quantifying accuracy loss versus performance gain for each approximation technique. However, this step has often been overlooked; while some works employ random error injection for kernel sensitivity, such methods fail to provide comprehensive insight. The final step is an efficient Design Space Exploration (DSE) process that rapidly explores approximate configurations to identify optimal or near-optimal trade-offs. Such heuristic algorithms are reviewed in Section 3.8.

1.4 The Advantages of Reconfigurable over ASIC Platforms

Choosing a suitable hardware platform for implementing an application depends on many factors, as illustrated in Fig. 1.2. Although ASIC devices offer excellent energy efficiency across a broad range of programs, reconfigurable fabrics provide far greater flexibility due to their post-fabrication datapath customization capabilities. This flexibility not only accommodates the rapid pace of algorithmic advancements relative to hardware update cycles but also supports the customizations required by different users and the ability to adapt to run-time workload variations [YYa+19, Gon+19, CMG15].

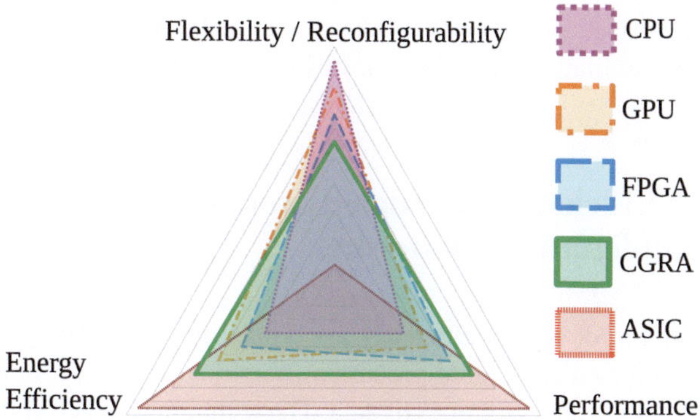

Fig. 1.2 Flexibility, performance, and energy-efficiency comparison between CGRA and other architectures [Liu+19b]

For example, state-of-the-art studies have demonstrated the benefits of CGRAs [Duc+19, Duc+17] in wearable systems, including commercial devices such as Samsung Galaxy smartphones and smartwatches [Kim+14, Ana15]. In addition, off-the-shelf FPGAs are commercially attractive due to their low Non-Recurring Engineering (NRE) cost, rapid prototyping capability, and short time-to-market. The high degree of parallelism and throughput offered by FPGAs and CGRAs makes them compelling choices for implementing hardware accelerators. In fact, many FPGAs incorporate dedicated DSPs blocks and soft Intellectual Propertys (IPs) to accelerate multiplication operations efficiently.

Reconfigurable fabrics—particularly FPGAs—are increasingly deployed across the edge-to-cloud continuum, spanning embedded edge platforms, networking devices, and cloud/data center servers. At the edge, AMD/Xilinx offers Artificial Intelligence (AI) accelerators designed to implement a variety of ML models on embedded devices [Fu+17b]. IBM has similarly developed a CGRA, named RaPiD, for accelerating AI workloads [Ven+21b].

Beyond edge devices, major telecommunications companies and cloud providers make extensive use of FPGAs in switches, routers, base stations, Network Inference Cards (NICs), and Network Processing Units (NPUs) [Bob+22, Wan+22b, Jin+20, Joh+20, Nie+20, Cis22, Xil23, Xil20, Adv24b, Int19, Int23a, Int21, Intar, Nap18, AMD22, Fir+18]. For instance, FPGAs are used as co-processors to accelerate tasks such as control and traffic management, load balancing, statistics co-processing, and cryptography/security analysis [Iba+19, Bre+20]. Companies including Microsoft, Amazon, Intel, and Huawei deploy FPGAs extensively in their data centers and cloud servers [Put+14, Ama18, Int23b, Xil17b].

The choice between FPGAs and CGRAs depends on several factors, particularly the type of parallelism—bit-level or word-level—required by the target application. CGRAs are generally more suitable for applications where the distribution of specific ALU operations (e.g., addition, multiplication, and division) is more pronounced. In such cases, CGRAs can often achieve higher performance and lower area/power consumption relative to FPGAs [Liu+19b]. In Sect. 2.2 of Chap. 2, we compare the performance metrics of individual ALU operations when implemented using Look-up Table (LUT) -based designs on FPGAs versus ASIC-based designs on CGRAs, along with their cumulative energy use in various applications.

1.5 Energy-Efficiency and High-Throughput Techniques for Reconfigurable Platforms

High throughput and/or energy efficiency are crucial for accelerating parallelizable applications—such as many signal processing workloads and ML models—which are frequently fed with large volumes of data. The techniques aimed at improving the throughput and energy efficiency of reconfigurable systems are summarized below.

Dedicated DSP Units Many FPGAs, including those from AMD/Xilinx and Intel/Altera, exploit a high degree of parallelism by incorporating hard-wired DSP blocks to accelerate multiplication [Xil18a, Cor20a]. While these blocks provide high performance, their usage may not always lead to the most optimal trade-offs in terms of overall performance and resource utilization across all applications [Ull+18, Ull+21, KR07]. Furthermore, despite their advantages, DSPs often fail to meet design requirements due to: (1) Their low ratio to LUTs (typically $<$ 0.001), which can be insufficient for concurrency-heavy or multiplication-intensive applications. While manual floorplanning may help for small designs, it is not feasible for complex applications competing for multiple FPGA resources. (2) Their fixed physical locations on the device, which introduce routing overhead and can degrade the performance of certain circuits [KR07], as shown for Viterbi decoders, Reed-Solomon, and JPEG encoders [Ull+21, Ull+18, KR07]. (3) Their inefficiency when performing multiplications at precisions smaller than 18×18 bits [BYB18, SLe+19], which prevents achieving the expected gains when precision-scaling techniques are employed.

Soft IPs To address limitations associated with DSPs, designers often resort to soft IP implementations of multipliers and dividers provided by major FPGA vendors such as Xilinx and Intel [Xil15, Xil16b, Cor14]. Using soft IPs for low bit-width operations is also recommended by both industry and academia [Xil18a, VL20]. However, the long latency and high resource overhead of LUT-based IPs may still restrict their applicability in resource- and energy-constrained environments such as wearable devices or aerial platforms. Additionally, the quest for high throughput remains unaddressed, motivating further improvement strategies summarized next.

Pipelining To increase throughput despite stagnating clock frequencies, pipelined versions of arithmetic IPs have been introduced by FPGA vendors such as AMD/Xilinx and Intel/Altera. However, none of the existing approximate dividers [YHS20, VL20, VJ19, Ull+18] have been designed specifically for LUT-based architectures or optimized for pipelined implementations. Indeed, pipelining has mostly been applied at coarse granularities within processor datapaths, whereas fine-grained pipelining—especially in approximate arithmetic—has been overlooked. Addressing this gap, RAPID designs are introduced in Chap. 4 to enable fine-grained pipelining of inexact multiplication and division specifically for FPGAs.

Parallelization through SIMD and MIMD Execution Models Many signal processing and ML applications rely heavily on matrix multiplication and convolution, making them well-suited for Single Instruction, Multiple Data (SIMD) execution. The rapid growth of error-resilient workloads, combined with the demand for high throughput, has motivated the development of imprecise SIMD components for FPGAs. To address this need, the concept of SIMD has been adopted by Xilinx [Fu+17a], Intel [LGB20, MLa+19], and IBM [Ven+21b, Ven+20] to: (1) support precision variability *within a single unit*, (2) exploit data-level parallelism inherent in applications, and (3) reduce routing and control overhead relative to logic.

However, these efforts have been largely limited to addition and multiplication with a common operand (including in CGRAs [Ven+21b, Ven+20]). While addition and multiplication are common operations, previous studies [Bra+18, HHS19] and our measurements (Figs. 2.4 and 2.5) show that divider latency—typically $2\times$ to $8\times$ that of a same-width multiplier—not only restricts application throughput[4] but also occupies a substantial share of ALU area and energy. Consequently, these challenges have prevented division-inclusive kernels from being offloaded to most CGRA accelerators , forcing the host processor to handle divisions and increasing communication and context-switching overhead.

Furthermore, the precision required for SIMD-packed operations is not uniform across or even within applications. For example, different layers of a single NN often require different precisions [Jud+16, Jud+15]. To address these challenges, the SIMDive multiplier-divider (Chap. 4) is proposed—customized for both FPGA [EUK20] and CGRA platforms [EK21]. SIMDive is the first hybrid multiplier-divider in the literature that supports both SIMD and Multiple Instruction, Multiple Data (MIMD) execution for multiplication and division operations.

1.6 In-Network Computing (INC) Paradigm

As discussed earlier, approximate computing has emerged as a powerful paradigm for achieving high throughput and energy efficiency by relaxing strict accuracy constraints. Its adoption has so far been most prominent in edge devices, where resource efficiency is critical, and in cloud infrastructures, where approximation can reduce service costs and improve response times. However, as data-driven applications continue to grow in scale and complexity, the traditional edge-cloud model struggles to satisfy the combined requirements of real-time responsiveness and energy efficiency. Offloading computation from edge devices to centralized clouds introduces substantial transfer overheads, latency, and energy consumption. These issues are amplified in the context of 5G and upcoming 6G networks, where applications such as video conferencing, real-time speech recognition, and e-healthcare must process massive volumes of data under tight latency, throughput, and reliability constraints.

To address these challenges, computing is increasingly shifting beyond the binary edge-cloud model toward decentralized and distributed configurations, where processing is performed closer to the data sources or intermediate network nodes. In fact, cloud-only approaches are becoming less sustainable for real-time, data-intensive workloads, due to the following limitations:

(1) *Response Time*: Data transmission constitutes a major portion of the total execution time in many cloud-based applications (e.g., more than 70% in some Facebook MapReduce jobs [Cho+11]). The projected increase in network traffic in 6G will only exacerbate latency in such applications.

[4] Division is often the slowest arithmetic operation in soft processors.

(2) *Processing Speed*: *General-purpose* processors in the cloud (typically up to 50 Gbps in Amazon) operate far slower than programmable switches (12–50 Tbps) [KT23], making cloud offloading insufficient for rapid, time-critical decision-making.
(3) *Energy Efficiency and Sustainability*: A large fraction of energy—up to 50% in many scenarios—is spent solely on transmitting data from endpoints to cloud servers [Cis20a], contributing significantly to carbon emissions. Moreover, the electricity cost required to *run* cloud services has already surpassed the cost of purchasing hardware for entire data centers [Hu+21].

These limitations motivate processing data "on the fly" as it traverses the network, a concept known as INC [Sap+17].

INC has emerged as a promising paradigm to mitigate the drawbacks of centralized edge-cloud solutions by leveraging a chain of network devices (e.g., switches, routers, NICs) to perform computation. Although still in its early stages, initial studies [Mic23, Zhe+23b, KT23] demonstrate substantial benefits when offloading *single-kernel* applications into the network. For instance, by moving computation closer to the data, INC has been shown to reduce *network traffic* by up to 90% for specific tasks [Sap+17]. Additionally, applications have shown improved *response time* and *energy savings* compared to centralized cloud execution [Tok+19, Jia+23]. Since intermediate switches are already part of the communication pipeline, most of these costs are inherently absorbed by packet processing and forwarding.

However, while these developments are encouraging, current INC research and deployments remain limited to relatively simple, single-kernel applications. Extending INC to multi-kernel workloads is substantially more challenging due to two major reasons: (1) Programmable network devices such as ASIC switches have limited memory and computing resources [Int22]. (2) These switches still lack support for FP operations and complex arithmetic functions [XZ19, MSV22], which are necessary for many advanced applications.

Although these limitations are less restrictive in FPGA-based network elements, the performance gap between ASIC- and FPGA-based implementations for *certain* applications may still limit achievable gains from in-network acceleration. Consequently, applying INC to complex tasks—such as running AI/ML inference inside the network—remains an open challenge in 5G and future 6G systems [Zhe+22a, Beh+20]. Fully unlocking the benefits of INC therefore requires rethinking the *computing* approaches used to execute multi-kernel applications.

To address these challenges within INC, prior research has applied approximation techniques either to *data type representation* (e.g., converting FP to fixed-point or integer [Pat+22, XZ19]) or to specific *primitives and functions* (e.g., multiplication [Sha+17], logarithmic operations [DSS20], matrix multiplication and division [Zhe+22b, Zhe+22a, Zhe+23b], and trigonometric functions [Kun+21]). Despite the promise shown by these works, their techniques remain limited to approximating *single* or *small-sized* kernels (e.g., edge-detection filters [Gle+19] or quantizing specific NN layers [SB18, SSB18, Luo+22, Lui+21, Saq+21]) and rely on *one* particular approximation technique rather than combining multiple ones.

1.7 Research Challenges

Despite technological and architectural advancements in reconfigurable computing, such systems still consume more resources and yield lower operating speeds compared to their ASIC-based counterparts. To address these limitations, approximate computing has emerged as a promising approach for reducing resource usage, latency, and energy consumption in error-resilient applications accelerated on reconfigurable platforms.

Although numerous *cross-layer* approximation strategies exist in the literature, the combination of techniques they employ does not typically form a cohesive cross-layer hierarchy. As a result, application-level optimizations may not be fully reflected in the underlying hardware—or can even be constrained by non-optimized circuits—and vice versa. Furthermore, most of these strategies focus solely on the *computing* phase of application execution while overlooking approximations in the data acquisition phase. Due to the limited performance gains achieved by such approaches, deploying multi-kernel applications on resource-constrained nodes remains challenging. Overcoming these limitations is especially critical in the 5G/6G era, where reducing the processing of large volumes of noisy and/or redundant data is essential for enabling an INC paradigm.

Overall, despite significant recent progress, reconfigurable devices still need further improvements in energy efficiency and throughput to be effectively deployed across the edge-to-cloud continuum and support increasingly data-intensive multi-kernel applications. Addressing these shortcomings remains an open research challenge. In the following, we outline the key challenges associated with this problem and map them to the relevant abstraction layers within a cross-layer approximation methodology.

1.7.1 Circuit-Level Challenges

Although considerable work has aimed to improve throughput and reduce energy consumption at the circuit level, several challenges remain. First, most existing approximate functional units are designed specifically for ASIC-based platforms. Because of the fundamental architectural differences between ASICs and reconfigurable fabrics, these techniques do not yield comparable performance improvements when synthesized and deployed directly on FPGAs.

Additionally, the majority of circuit-level approximation research has focused on addition and multiplication, while the high energy consumption and long latency of division have been largely underexplored. This has limited the gains achievable for division-inclusive applications. Among the few approximate dividers available in the literature, none has been tailored specifically for LUT-based architectures.

Moreover, most FPGA-targeted inexact multipliers rely on tree-based designs, approximating the accurate structure of multipliers that operate on a 2D array of PPs.

However, adhering to array-based structures is another factor limiting performance gains—particularly compared to logarithmic approaches.

Finally, all existing approximate multipliers and dividers operate exclusively in SISD or non-pipelined modes, making them unsuitable for high-throughput applications that process large data volumes or exhibit parallelism (e.g., matrix multiplication). Fine-grained pipelining and SIMD execution strategies within approximate arithmetic units have been largely overlooked; to date, these techniques have only been implemented at coarse granularities, such as in processor datapaths or CGRAs, where they appear through interconnected PEs.

1.7.2 Architecture-Level Challenges

As previously discussed, architecture-level techniques for improving throughput and energy efficiency often involve modifying or connecting Single Instruction, Single Data (SISD) adders and multipliers within PEs to support SIMD execution. However, implementing SIMD at the intra-PE level—where functional units operate in both SISD and SIMD modes—is resource-intensive. Conversely, inter-PE SIMD execution, which connects multiple PEs, introduces significant mapping and routing complexity and increases area, power usage, and latency in the CGRA.

Moreover, prior efforts to enable SIMD have focused primarily on adders and multipliers; extending SIMD to dividers is not straightforward, as constructing a higher-order *accurate* divider from smaller units using hierarchical composition is not mathematically viable. Consequently, the absence of SISD/SIMD division in existing approximate CGRAs has prevented offloading division-inclusive kernels, forcing the host processor to handle these computations and increasing context-switch overhead. Even heterogeneous architectures that incorporate limited imprecise SISD dividers suffer from reduced operating frequencies due to the inherently higher latency of division compared to multiplication. This limits achievable throughput when deploying applications on such reconfigurable devices.

Finally, because application kernels exhibit varying levels of error resilience to precision reduction, fixed-precision functional units in approximate CGRAs are unsuitable for efficiently executing all operations within a kernel or across multiple kernels. The absence of runtime precision adaptation leads to:

(1) resource underutilization, weakening the benefits of application-level approximation,
(2) poor performance for reduced-precision operations due to fixed ALU latency, and
(3) an inability to offload concurrent kernels to the available PEs. These limitations highlight the need for architectures capable of dynamically adapting circuit-level precision at runtime to reflect application-level approximation requirements.

1.7.3 Application-Level Challenges

Except for NNs, most existing approximation frameworks use uniform and/or fixed-precision strategies when assigning precision levels across different kernels of a multi-kernel application. However, fixed-precision approaches ignore the fact that kernels differ in their sensitivity to approximation—both in terms of *achievable performance gains* and *resulting QoR degradation*. This can lead to inefficient balances between performance and accuracy.

Determining the optimal adjustment of approximation knobs in multi-kernel applications remains challenging for two key reasons. First, without a comprehensive error-resilience sensitivity analysis—particularly in the presence of multiple approximation techniques—it is impossible to identify the relative saliency and prioritization of kernels. Such an analysis is essential for understanding performance-accuracy trade-offs but has been largely neglected in prior research. Second, tuning multiple approximation knobs across multiple kernels is a complex multi-objective optimization problem where the design space grows exponentially with the number of kernels and approximation parameters. Thus, an efficient DSE method is required to rapidly explore a vast configuration space and identify optimal or near-optimal solutions.

1.7.4 Cross-Layer Challenges

Current cross-layer approaches cannot effectively apply a chain of approximations within a cohesive cross-layer hierarchy. As discussed earlier, most techniques apply application-level approximations (e.g., quantization) on top of circuit-level approximations (e.g., approximate operators) while relying on fixed-precision hardware. This ignores optimization opportunities at the architectural level—such as SIMD, which could accelerate multiple reduced-precision operations in parallel—and can even cause application-level optimizations to be poorly realized by non-optimized hardware. For example, performing precision scaling on multipliers mapped to FPGA DSP blocks can underutilize these fixed hardwired units, limiting achievable gains in area and latency.

Furthermore, most cross-layer work has focused exclusively on NNs, with relatively few efforts targeting other application domains. The field lacks a general approximation methodology capable of efficiently tuning multiple approximation knobs across both the data acquisition and computing phases of an application, and across multiple abstraction layers in a multi-kernel context. A framework that can generate near-optimal configurations rapidly—maximizing performance while respecting user-defined accuracy thresholds—is highly desirable.

1.7.5 Challenges for INC

At present, adopting an INC approach for multi-kernel applications faces three key challenges. First, ASIC-based programmable switches do not support FP types or complex arithmetic operations such as multiplication and division [XZ19, MSV22]. Second, these devices have limited memory and logic resources [Int22], restricting their ability to execute computationally intensive tasks such as matrix multiplication or convolution used in many ML and signal-processing models. Although FPGA-based network elements mitigate some of these limitations, the performance gap between ASIC- and FPGA-based implementations for certain applications may still limit the benefits of in-network acceleration . Finally, orchestrating the distribution of application kernels across a chain of heterogeneous network nodes—while meeting end-to-end latency and QoR requirements—remains an open research challenge.

Overall, these unresolved issues have hindered the development of a systematic cross-layer methodology for reconfigurable fabrics that can fully exploit the synergistic benefits of multiple approximation techniques within a cohesive hierarchical structure. Consequently, this book focuses on addressing the following research problem:

Enabling an efficient cross-layer approximation methodology for reconfigurable systems through the design of high-throughput and energy-efficient functional units.

1.8 Book Key Contributions

Figure 1.3 illustrates the overall structure of this book. The objective is to develop a cross-layer approximation methodology aimed at achieving high throughput and energy efficiency for reconfigurable devices. In the following, the technical contributions of each chapter (that attempt to address the research challenges discussed in Sect. 1.7) are summarized.

1.8.1 Circuit-Level Contributions (Chap. 4)

As discussed earlier, multiplication and division are among the most frequently used and energy-intensive arithmetic operations across a wide range of applications, making them prime targets for approximation. To significantly reduce the design complexity of these operations, various logarithmic multipliers and dividers are developed using Mitchell's algorithms, which transform the 2D structure of array-based multiplication/division into 1D addition/subtraction in the logarithmic domain. This formulation is particularly suitable for FPGAs, as they incorporate fast carry chains optimized for addition/subtraction. A logarithmic design also greatly shortens the long latency of accurate division, bringing it closer to that of a same-size multiplier.

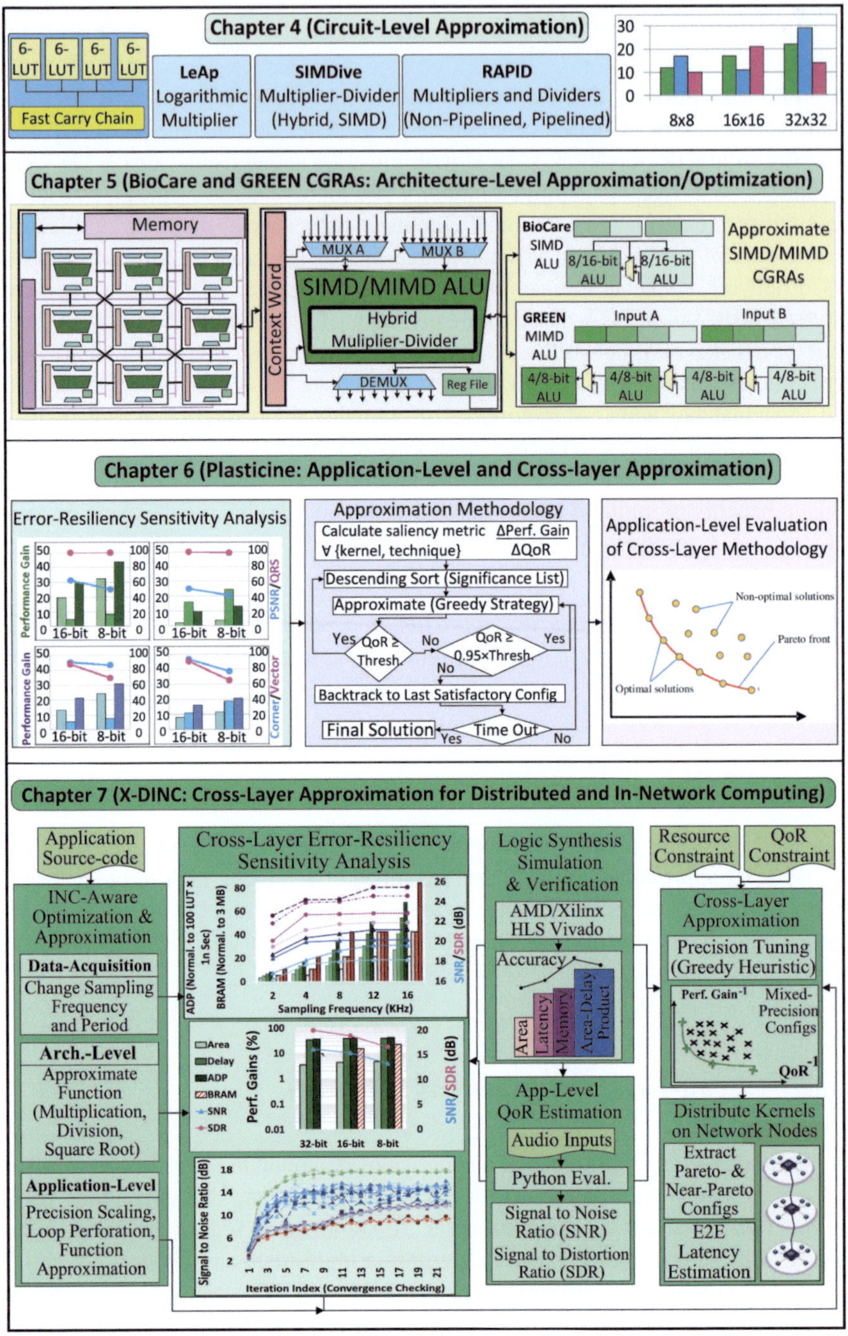

Fig. 1.3 Overall structure of the book

Furthermore, the proposed multipliers and dividers exhibit an Average of absolute Relative Error (ARE) below 1% and maintain near-zero bias, a critical property for minimizing accumulated errors in multi-kernel approximate applications. To target high throughput, several imprecise variants are also presented, incorporating fine-grained pipelining or SIMD/MIMD execution capabilities.

- The proposed logarithmic multipliers and dividers (LeAp in Sect. 4.2 and SIMDive in Sect. 4.4) support accuracy configurability. Their efficiency is evaluated both at the circuit and application levels, when deployed in various single- and multi-kernel programs from domains such as classification, audio, bio-signal, and image processing . Each design offers distinct features:
- SIMDive is the first integrated multiplier-divider that enables runtime switching between multiplication and division without requiring hardware reconfiguration. This hybrid functionality is especially advantageous for multiplication-intensive applications (e.g., ML or image processing) that nonetheless involve occasional division operations. Additionally, the methodology provides on-the-fly precision adaptability and supports an integrated hierarchy of approximation techniques—from application-level precision scaling to circuit-level imprecise multiplication/division—realized through an architecture-level SIMD/MIMD execution model.
- RAPID is the first approximate non-IP design enabling fine-grained pipelining within a single multiplier or divider. A range of RAPID configurations is provided, corresponding to different numbers of pipeline stages, enabling diverse power-throughput trade-offs and adaptability to varying operating frequency targets on FPGAs. Fine-grained pipelining at the intra-unit level can achieve higher frequencies than coarse-grained pipelining at the inter-unit level (e.g., within ALUs), a feature especially beneficial for multiplication-dominant workloads that process large batches of input data.

1.8.2 Architecture-Level Approximations/Optimizations (Chap. 5)

The proposed BioCare and GREEN architectures introduce area- and energy-efficient approximate CGRAs that support runtime functional versatility and precision adaptability through the plasticity of their lightweight PEs. Each PE can execute addition, multiplication, and division at multiple bit-widths. Importantly, enabling SIMD/MIMD execution at the *intra*-PE level (rather than the conventional inter-PE level) alleviates pressure on global routing. Additionally, providing division support within each PE eliminates the need for context-switching between the CGRA and host processor when accelerating kernels that include division. More specifically:

- BioCare exploits the synergistic chain of approximation and optimization techniques within a cohesive cross-layer hierarchy. Leveraging the SIMDive unit, each

PE supports precision scaling (application level) on top of imprecise multiplication and division (circuit level). Using a SIMD execution model, the BioCare CGRA can operate in either SISD mode (16-bit) or SIMD mode (8-bit subwords), selected at runtime.
- The GREEN CGRA extends BioCare by incorporating MIMD execution and supporting a broader range of heterogeneous operations (ADD-MUL-DIV opcodes) at 16-, 8-, and 4-bit precisions. In addition, memory and reconfiguration overheads are accounted for in GREEN's full architectural evaluation.

1.8.3 Application-Level and Cross-Layer Contributions (Chap. 6)

Although combining multiple approximation techniques may seem straightforward, determining an effective combination across the kernels of an application is far from trivial, especially due to error propagation. These factors motivate the need for a systematic approximation methodology that leverages the synergistic effects of multiple techniques while appropriately tuning optimization knobs to maximize performance and minimize accuracy loss.

To our knowledge, no prior work has proposed a generic cross-layer methodology that properly adjusts approximation and optimization knobs across a multi-tier hierarchy spanning data acquisition, architecture, and circuit layers. Such an adaptable methodology—capable of generating energy-efficient and high-throughput configurations—is highly desirable for applications deployed across the edge-to-cloud continuum.

To this end, this book introduces *Plasticine*, a cross-layer methodology that harnesses synergistic approximation effects across multiple abstraction layers. Plasticine enables a cohesive chain of approximations, from application-level precision scaling to architecture-level SIMD/MIMD execution and circuit-level imprecise multiplier-dividers. It dynamically tunes approximation knobs for multi-kernel applications to achieve user-defined QoR constraints. The methodology is evaluated on several widely used single- and multi-kernel applications from classification, biosignal, audio, and image-processing domains on both FPGA and CGRA platforms. Specifically, the following contributions are provided:

- An application-level error-resiliency sensitivity analysis that determines the relationship between end-to-end *performance gain* and *QoR degradation* for individual kernels subjected to different approximation techniques.
- A greedy-based DSE heuristic that rapidly tunes cross-layer approximation knobs, avoiding exhaustive search. The resulting Pareto or near-Pareto *mixed-precision* configurations offer diverse performance-accuracy trade-offs and outperform accurate or uniform-precision counterparts across a wide range of accuracy levels.

1.8.4 Cross-Layer Approximation Methodology for Distributed and In-Network Computing (Chap. 7)

To fully unlock the potential of INC and support larger applications benefiting from in-network acceleration, the *computing* model for executing multi-kernel applications must be rethought through the lens of approximate computing. To achieve this, Chap. 7 presents *X-DINC*, an extended and customized version of the Plasticine cross-layer methodology tailored for distributed and INC environments.

Compared to accurate or centralized execution, X-DINC substantially accelerates multi-kernel applications while reducing resource footprint—even when kernels are distributed across multiple network nodes. Specifically, X-DINC provides the following features:

- X-DINC restructures the computational graph of the application to suit an INC setting by approximating the implementation of complex arithmetic operations that are unsupported on programmable network switches.
- X-DINC identifies chains of approximation and optimization opportunities across both the data acquisition and computation phases when the application is executed within an INC environment.
- A broadened sensitivity analysis and a greedy heuristic algorithm, respectively, evaluate the error resilience of individual kernels and tune approximation knobs across consecutive kernels to maximize performance gains under user-defined QoR constraints.
- Finally, based on user-specified accuracy thresholds or available node count, X-DINC selects an appropriate configuration from the set of optimal or near-optimal candidates and distributes the kernels across a chain of network nodes, minimizing transmission latency and achieving the lowest end-to-end response time.

1.9 Book Overview

The rest of this book is organized as follows. Chapter 2 presents the background knowledge and preliminaries, required for the book. Chapter 3 provides a summary of state of the art approximation techniques, applied at different layers of abstraction. Moreover, sensitivity analysis approaches, cross-layer methodologies, and DSE strategies are discussed. An in-depth look into the approximation techniques for the emerging paradigm of INC is also presented in Chap. 3. The key contributions of this book are detailed in Chaps. 4 (circuit-level), 5 (architecture-level), 6 (application-level and cross-layer), and 7 (customization and expansion of cross-layer methodology for a distributed and INC environment). Finally, Chap. 8 concludes the book and provides insights into the potential future tracks.

Chapter 2
Background and Preliminaries

Abstract This chapter provides the preliminary information and related work that are required to understand the various contributions of the book better. Specifically, it provides information about the architecture of Xilinx FPGAs in Sect. 2.1. The structure of array-based accurate multiplication and division is demonstrated in Sect. 2.2. Afterward, Sect. 2.3 compares the performance metrics of accurate addition, multiplication, division, and square root operations at different bit-width precisions and their contribution to the ALU energy breakdown across various single- and multi-kernel applications . Mitchell's multiplication and division approximate algorithms used to design the functional units in this book are discussed in Sect. 2.4. Section 2.5 introduces and discusses the various standard statistical error metrics used to characterize the designed approximate operators. Finally, Sect. 2.6 briefly introduces and demonstrates the structure of the multi-kernel applications that have been used in the experiments in the book.

Keywords FPGA · Mitchell's algorithm · Logarithmic hybrid multiplier-divider · Blind source separation · Image processing · Bio-signal processing · Acoustic anomaly detection · Neural networks · Independent component analysis · Error metric

2.1 Xilinx FPGA Slice Structure

Cutting-edge FPGAs, provided by Xilinx and Intel, utilize 6-LUTs to implement various circuits [Xil16a, Cor20b]. In this book, the Xilinx FPGAs is used to implement all designs. The reason is that Xilinx FPGA device families can better exploit the simple shift-and-add approach [Xil15, Xil13], whereas Intel's can efficiently implement Booth algorithm [Cor14].

The Configurable Logic Block (CLB) is the main computational block of FPGAs for implementing various circuits. The CLB of a modern Xilinx FPGA, such as Xilinx UltraScale, consists of a slice of eight 6-LUTs, 8-bit carry chain and 16 Flip Flops (FFs) [Xil17a]. The same resources are arranged in two slices of a Xilinx 7 series FPGA [Xil16a] (used for the implementations in this book). As shown in Fig. 2.1a,

Z. Ebrahimi and A. Kumar, *Cross-Layer Approximation and In-Network Acceleration*,
https://doi.org/10.1007/978-3-032-21712-7_2

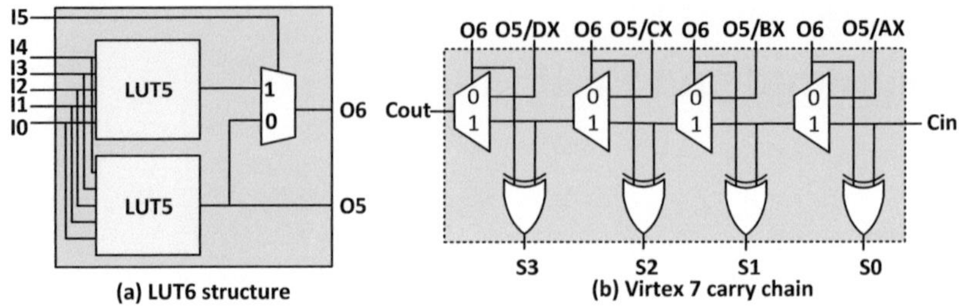

Fig. 2.1 The structure of Xilinx FPGA slice [Xil16a]

a 6-LUT can be used to implement either a single 6-input combinational circuit or two 5-input circuits. For the configuration of 6-LUT, a 64-bit INIT value is assigned to it. This INIT value denotes all the input combinations of 6-LUT for which a true value (1) is received at the output. For example, an INIT value of 0000000000000002 (hex) for 6-LUT defines to produce outputs O5 = 1 and O6 = 0 for input combination 100001. Besides the implementation of combinational functions, these 6-LUTs are also used for controlling the associated carry chain, as shown in Fig. 2.1b. The carry chain implements a Carry Look Ahead (CLA) adder using O5 as the carry-generate signal and O6 as the carry-propagate signal as described by Eqs. 2.1 and 2.2. However, O5 can be bypassed by the external IX signal to provide the carry-generate signal. The input carry, "CIN," can be either assigned to constant '0/1' or to "COUT" of another carry chain from a different slice.

$$S_i = P_i \oplus C_i \tag{2.1}$$

$$C_{i+1} = G_i.\bar{P_i} + P_i \cdot C_i. \tag{2.2}$$

2.2 Accurate Multiplication and Division Algorithms

2.2.1 Accurate Multiplication

Figure 2.2 illustrates various accurate multiplication methods, including Wallace [Wal64], Dadda [Dad65], Baugh-Wooley [BW73], and Booth [Boo51]. Table 2.1 also summarizes the comparison of these methods. These methods first generate a 2D array of PPs according to predefined formulation (which consists of performing logical *AND* operations on all the bits of both operands) and then add them, using compressors/adders to compute the final product. The 3:2 and 2:2 compressors, shown in Fig. 2.2, are used for the reduction of PPs and correspond to FAs and HAs, respectively. While Wallace and Dadda can handle unsigned numbers, Baugh-

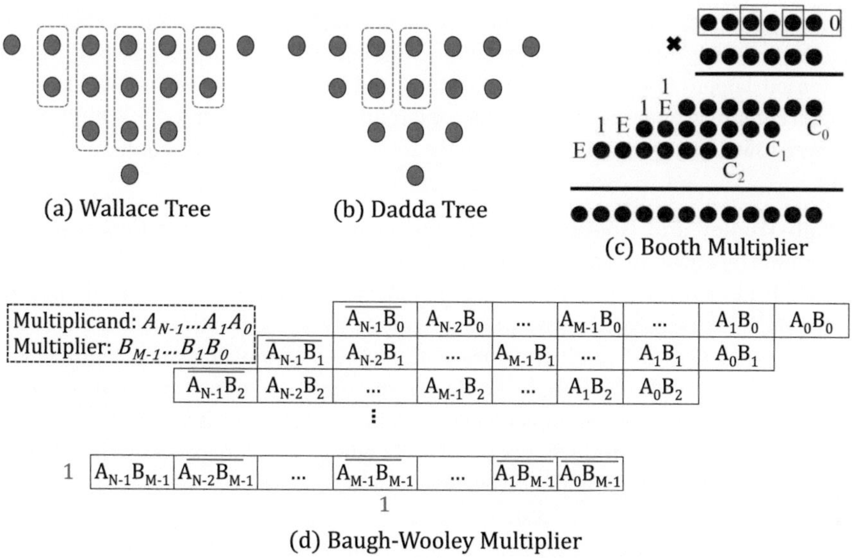

Fig. 2.2 The structure of accurate multiplication with different approaches

Table 2.1 Comparison of Wallace, Dadda, Baugh-Wooley, and Booth multiplication approaches

Approach	Latency	Area	Signed handling	PP addition/reduction
Wallace tree	Low	High	Requires additional logic	3:2 and 2:2 Compressors
Dadda tree	Moderate	Moderate	Requires additional logic	3:2 and 2:2 Compressors
Baugh-Wooley	High	Moderate-high	Built-in for two's complement	Carry propagation adder
Booth encoding	Moderate-high	Low-moderate	Built-in with encoding	Carry-save adders

Wooley and Booth are tailored for signed numbers by eliminating the need for extra sign-extension bits. It should be noted that among other available multiplier design options, the conventional shift-and-add [Par10], serial, and serial/parallel multipliers address the low area requirements but offer high critical path delays.

Wallace and Dadda Both approaches use a tree-based structure and are commonly employed techniques for computing the final products using compressors. Both also have relatively high area requirements to achieve low output latencies by parallelizing the addition of PPs. However, they differ in how they reduce the PPs. While Wallace is targeted for simplicity and high speed, Dadda employs a more complex tree structure aimed at minimizing the number of PPs and adders, thereby improving resource efficiency. Wallace uses as many parallel 3:2 compressors as possible, reducing the number of reduction stages but requiring more complex wiring. On the other hand, Dadda minimizes the number of reduction stages, leading to reduced area but slightly increased latency. Figure 2.2 also illustrates the initial stages of the Wallace and Dadda compression trees, which reduce the four rows of PPs (for a 4×4 multiplier) to three rows. The rectangular boxes in the figure indicate the utilization of HAs and FAs.

Baugh-Wooley and Booth Both methods handle the sign information in the PP generation by encoding consecutive bits of the multiplier into 2-bit Booth codes, i.e., 00 (0), 01 (+1), or 11 (-1). This encoding reduces the number of generated PPs and adder stages compared to conventional methods. Overall, Booth encoding enables fast multiplication, although it comes with the cost of higher resource utilization than tree-based methods. Baugh-Wooley is optimized for signed multiplication by directly supporting two's complement numbers. On the other hand, Booth reduces the number of partial products via bit-pair encoding, which is particularly efficient for multiplying numbers with consecutive 1s (since it reduces switching activity), but it requires encoding and decoding steps.

The aforementioned multiplication approaches have been adopted by many researchers to design various multipliers customized for ASIC and FPGAs. For example, Kumm et al. have utilized Integer Linear Programming (ILP) to explore the modular design for resource-efficient implementation of larger multipliers from smaller sub-multipliers in Xilinx FPGAs [Kum+17]. Their technique considers both the DSP units and LUTs to implement the sub-multipliers to efficiently utilize the available resources on the FPGA. The authors of [Far+20] have utilized the 6-LUTs and carry chains of Xilinx FPGAs to implement a set of constant-coefficient multipliers for NNs . These multipliers are based on the shift, add, and subtract operations. Parandeh-Afshar et al. have also used Booth's and Baugh-Wooley's multiplication algorithms [BW73] for area-efficient multiplier implementation using Altera (now Intel) FPGAs [PI11]. However, to reduce the effective length of carry chains, their implementation limits the length of the Adaptive Logic Modules (ALMs) to five, which results in the under-utilization of the FPGA resources. Moreover, this feature of limiting the carry chain to five ALMs cannot be achieved without wasting resources, with current FPGAs from other vendors, such as those provided by Xilinx [Xil16a]. The authors of [LB18] have utilized the ALMs of Intel FPGA to propose resource-efficient implementations for smaller multipliers, such as 3×3 and 5×5. Their implementation focuses on restructuring the partial product terms to reduce the total number of levels required to compute the final product.

2.2.2 Accurate Division

The array divider , illustrated in Fig. 2.3, uses an array of subtraction modules to perform division through a series of iterative subtractions and comparisons. The array divider is particularly useful in digital circuits and can be implemented using basic logic gates and subtractors. In the 2D structure of an array divider (illustrated in Fig. 2.3), each subtraction module performs the operation of subtracting the divisor from a portion of the dividend (The division operation $\frac{a}{b} = c$ involves the dividend a, the divisor b, and the quotient c). The iterative shift-and-subtract process of accurate division is as follows. For each bit position (starting from the most significant bit): (1) Shift the partial remainder one bit to the left to make space for the next bit of the dividend. (2) Subtract the divisor from the current value in the partial remainder. (3) Check the result of the subtraction: If the result is non-negative, set the corresponding bit in the Quotient to 1. If the result is negative, restore the previous value of the partial remainder (undo the subtraction) and set the corresponding bit in the Quotient to 0. This process is repeated for all bit positions of the dividend. The main difference between restoring and non-restoring divisions lies in their remainder handling: Unlike restoring division, which fixes negative remainders immediately, the non-restoring version Keeps the negative remainder (does not restore) and compensates by adding the divisor in the next cycle. Non-restoring is generally preferred in modern designs because it is faster (no restore operations), has a more efficient hardware implementation, and renders the same accuracy to restoring version after final correction [ESS20, SES19, LCh+18, LCh+16]. It is important to note that constructing a higher-order divider from smaller instances through a hierarchical approach is not mathematically feasible.

2.3 Comparing the Performance Metrics of Accurate Operations in Different Bit-Width

Figures 2.4 and 2.5 compare the area, delay, and energy of different operations in various bit-width precisions in ASIC- and FPGA-based implementations, respectively. As can be seen in these figures, the area, latency, and energy of multiplication, division, and square root operations are significantly higher than the same-sized adder and multiplier (similar observations have been partly concluded in studies like [HHS19, HHS18a]). This motivates the use of approximate configurations for these more complex operations (circuit-level technique). The approximate components presented in this book are designed based on the logarithmic approach of Mitchell's multiplication and division algorithms.

ALU Energy Breakdown in Single- and Multi-kernel Applications

In Fig. 2.6 further demonstrates the energy consumption of different ALU operations (based on their distribution) in eight single- and multi-kernel applications. Our case-study applications cover tasks from pre-processing (band-pass Finite Impulse Response (FIR) filter for noise removal) to feature extraction (Fast Fourier Transform (FFT), Discrete Wavelet Transform (DWT), and approximate entropy) and classification (K-Means). These multi-kernel applications are ubiquitous in classification, bio-signal (Electrocardiogram (ECG)/Electroencephalogram (EEG)), image-, and video-processing domains. For example, the Pan-Tompkins algorithm is not only the

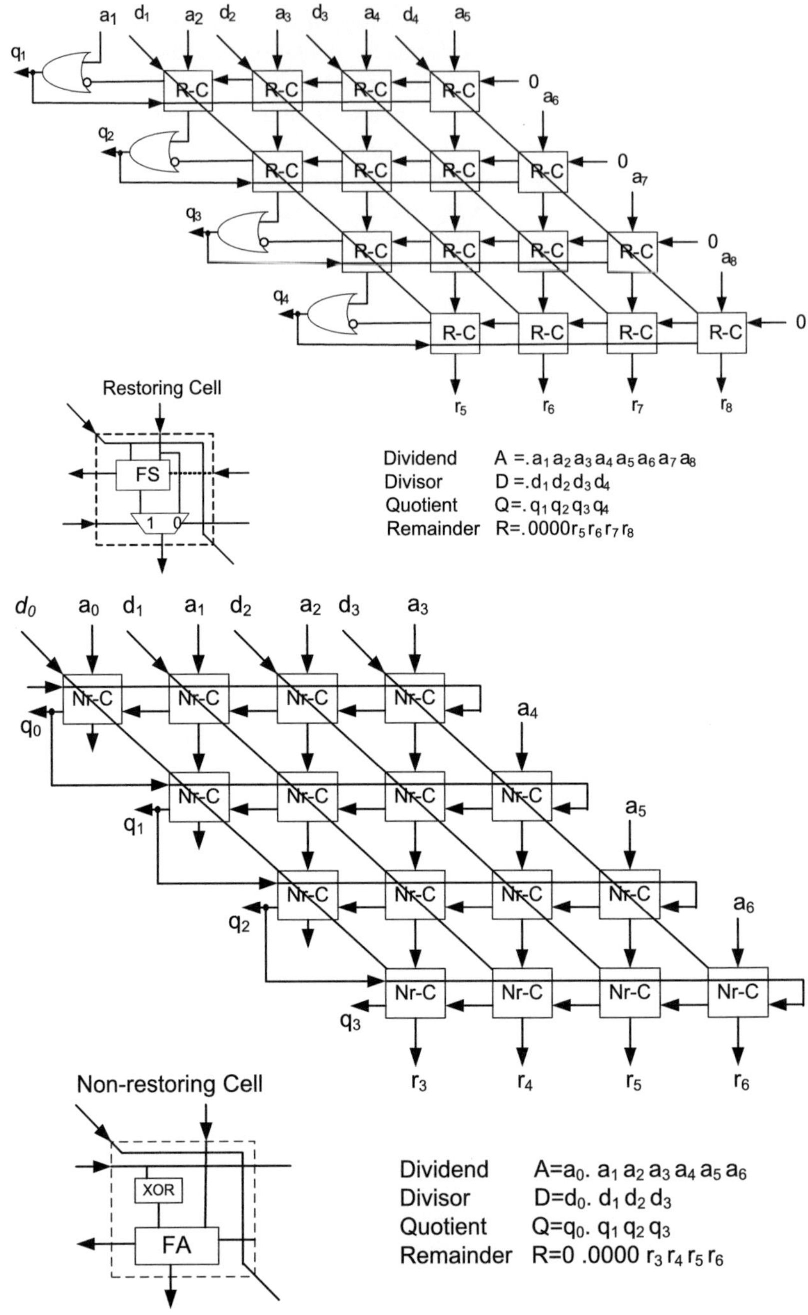

Fig. 2.3 Diagram of a 4-bit restoring (top) and non-restoring (bottom) array divider [Say+15]

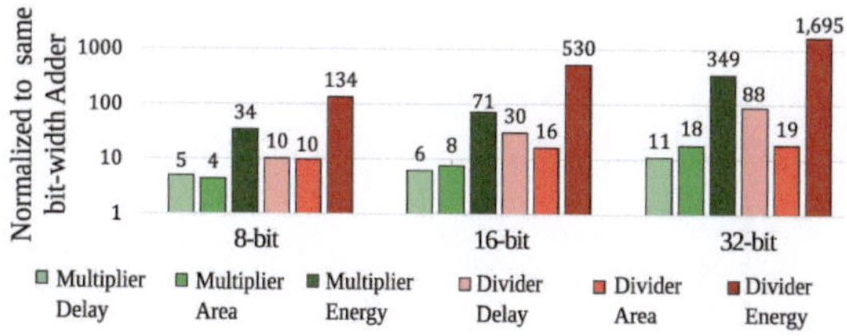

Fig. 2.4 Comparing area, delay, and energy of 8-, 16-, and 32-bit addition, multiplication, and division operations when implemented in an ASIC approach

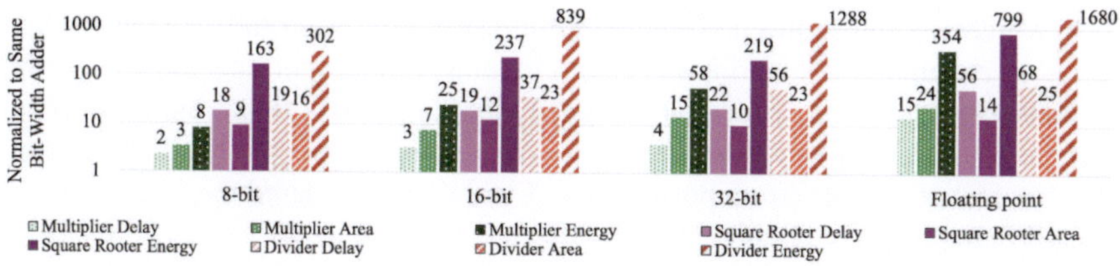

Fig. 2.5 Comparing area, delay, and energy of 8-, 16-, and 32-bit addition, multiplication, division, and square root operations, when implemented in Virtex-7 FPGA, via LUT-based accurate IPs

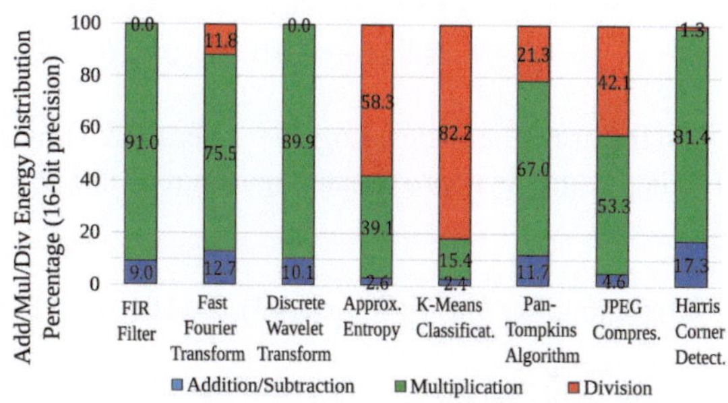

Fig. 2.6 The cumulative energy of ALU operations in bio-signal and image processing applications (16-bit precision)

atomic task in heart disease diagnosis or brain epilepsy and sleep apnea but also in biometric authentication. The results shown in Fig. 2.6 exhibit that although division is less frequent, it still consumes a substantial portion of ALU energy in the kernels of these applications.

2.4 Mitchell's Logarithmic Multiplication and Division Algorithms

As shown in Eq. 2.3, Mitchell has proposed approximate variations of the accurate multiplication and division operations to reduce the resource and delay overheads of the accurate unsigned multiplier and divider. As hinted in this Equation, Mitchell's algorithms transform the 2D structure of such array-based multiplier (divider) into a 1D adder (subtractor) in the logarithmic representation. Therefore, this transformation significantly reduces the complexity of the accurate designs.

$$
\begin{cases}
P = A \times B \xrightarrow[\text{Log}]{\text{Approx.}} \widetilde{\text{Log}_P} = \widetilde{\text{Log}_A} + \widetilde{\text{Log}_B} \xrightarrow[\text{Anti-Log}]{\text{Approx.}} \tilde{P} = 2^{\widetilde{\text{Log}_P}} \\
D = A \div B \xrightarrow[\text{Log}]{\text{Approx.}} \widetilde{\text{Log}_D} = \widetilde{\text{Log}_A} - \widetilde{\text{Log}_B} \xrightarrow[\text{Anti-Log}]{\text{Approx.}} \tilde{D} = 2^{\widetilde{\text{Log}_D}}
\end{cases}
\tag{2.3}
$$

As Mitchell's algorithms perform logarithmic multiplication and division, the first step is to calculate the logarithmic representation of the input operands. Consider the binary representation for N-bit unsigned input operand A (also similar for input operand B), which can be written as Eq. 2.4, where k reveals the position of the leading one. The rest of the bits (from position $k-1$ to 0) are considered the fractional part and fall in the $0 \leq x < 1$ range.

$$
\begin{gathered}
\text{Input} = 2^k + \sum_{i=0}^{k-1} 2^i b_i = 2^k(1+x) \xrightarrow{\text{Apply on input operand } A=43 \text{ and } B=10} \\
43 = 2^5(1+0.01011)_2, \quad 10 = 2^3(1+0.01)_2
\end{gathered}
\tag{2.4}
$$

In linear mathematics, $\text{Log}(1+x)$ is approximated to x for this range of $0 \leq x < 1$[1]; therefore, the approximate logarithm of inputs are:

$$
\text{Log}(\text{Input}) \simeq k + x \rightarrow \text{Log}(43) \simeq (101.01011)_2, \quad \text{Log}(10) \simeq (11.01)_2. \tag{2.5}
$$

After applying the same step on the second input to get its approximate Log, the summation (subtraction) of two parts is obtained in Eq. 2.6 (Eq. 2.7).

$$
\widetilde{\text{Log}(\tilde{P})} = (k_A + k_B) + (x_A + x_B) \rightarrow K_{\text{sum}} = (1000)_2, X_{\text{sum}} = (0.10011)_2 \tag{2.6}
$$

$$
\widetilde{\text{Log}(\tilde{D})} = (k_A - k_B) + (x_A - x_B) \rightarrow K_{\text{sub}} = (10)_2, X_{\text{sub}} = (0.00011)_2. \tag{2.7}
$$

Finally, by adding the error-reduction coefficient (optional step) and applying anti-log (a shift operation), the binary representation of the approximate product (quotient) is derived by Eq. 2.8 (Eq. 2.9):

[1] Throughout this book, Log stands for Log_2.

$$\tilde{P} = \begin{cases} 2^{k_A+k_B}(1 + x_A + x_B), & x_A + x_B < 1 \\ 2^{k_A+k_B+1}(x_A + x_B), & x_A + x_B \geq 1 \end{cases} \rightarrow \begin{cases} \tilde{P} = (110011000)_2 = 409 \\ P_{\text{accurate}} = 430 \end{cases} \tag{2.8}$$

$$\tilde{D} = \begin{cases} 2^{k_A-k_B-1}(2 + x_A - x_B), & x_A - x_B < 0 \\ 2^{k_A-k_B}(1 + x_A - x_B), & x_A - x_B \geq 0 \end{cases} \rightarrow \begin{cases} \tilde{D} = (100)_2 = 4 \\ D_{\text{accurate}} = 4 \end{cases} . \tag{2.9}$$

Mitchell-Based Approximation of Square Root
The square root operation (used e.g., in the structure of Blind Source Separation (BSS) application) is implemented via a non-restoring algorithm [Adv24a] (due to being resource-friendly), wherein the accurate multiplication is replaced with the Mitchell-based version.

2.5 Statistical Error Metrics

Statistical error metrics at circuit-level are used to evaluate the output quality of approximate circuits (O_{app}), compared to the accurate output (O_{acc}). This book employs some commonly utilized error metrics to characterize the proposed approximate circuits [Sta+20, SRe+16, HJi+20, SIS15]. An overview of these metrics are provided herein.

ARE: As defined in Eq. 2.10, this metric characterizes the absolute error magnitude with respect to the accurate answer and then averages these errors.

$$\text{ARE} = \frac{1}{N} \sum_{i=0}^{N} \left| \frac{O_{\text{Acc}} - O_{\text{App}}}{O_{\text{Acc}}} \right| . \tag{2.10}$$

Peak of absolute Relative Error (PRE): As defined in Eq. 2.11, this metric computes the peak of absolute relative error for all input combinations. Together with the average absolute relative error, this metric helps estimate the overall error distribution of an approximate operation. One should expect a few output values with high absolute relative error values.

$$\text{PRE} = \forall_{\text{inputs}} \max \left(\left| \frac{O_{\text{Acc}} - O_{\text{App}}}{O_{\text{Acc}}} \right| \right) . \tag{2.11}$$

Error Bias (EB): As defined in Eq. 2.12, this metric shows how errors are distributed around zero (i.e., have negative or positive signs). Several studies, including ours, have shown that a low error-bias error plays a pivotal role in minimizing the error accumulation in applications having multiple kernels with an aggregation-based structure (i.e., mainly addition/multiplication operations) [Zer+21, SAn+20, Tas+20, EK24, Ebr+22, Ebr+21].

$$\text{Error Bias} = \sum_{i=0}^{N} \frac{O_{\text{Acc}} - O_{\text{App}}}{O_{\text{Acc}}}. \tag{2.12}$$

Signal-to-Noise Ratio (SNR): As defined in Eq. 2.13, this metric measures the strength of the desired audio signal relative to the background noise. A higher SNR indicates a clearer and cleaner audio signal with less background noise. SNR is typically expressed in decibels (dB). In this formula, S and $\hat{S}$ denote the original and estimated source, respectively.

$$\text{SNR} = \sum_{i=1}^{n} 20 \log 10 \frac{|S_i|}{|\hat{S}_i - S_i|}, \quad i[1, 2, \ldots, m]. \tag{2.13}$$

SDR: As defined in Eq. 2.14, this metric, also used in audio processing , quantifies the quality of a signal that has been subject to processing, such as compression, enhancement, or separation. SDR measures the ratio between the power of the desired signal and the power of the distortion introduced by the processing, including noise, interference, and artifacts. In this formula, S and $\hat{S}$ denote the original signal, while e_{interf}, e_{noise}, and e_{artif} are the errors stemming from interference, noise, and artifacts, respectively. As can be seen, SDR measures the power ratio of the signal to all types of distortions, including noise, artifacts, and any other unwanted components introduced during processing. In contrast, SNR only measures the power ratio of the signal to the background noise. Moreover, SDR is commonly used in evaluating the performance of audio processing applications where various types of distortions are present. On the other hand, SNR is typically used in scenarios where the main concern is the noise level relative to the signal, such as in audio recording and broadcasting.

$$\text{SDR} = 10 \times \log 10 \frac{||S^2||}{||e_{\text{interf}} + e_{\text{noise}} + e_{\text{artif}}||^2}. \tag{2.14}$$

PSNR: As defined in Eq. 2.15, this metric measures the quality of a reconstructed or compressed signal compared to its original form. It is commonly used in image processing applications to assess the fidelity of the reconstructed image. Nevertheless, this metric is also used in other signal-processing domains, including biosignal (heart). In this formula, Max is the maximum possible pixel value of the image/signal. At the same time, MSE is the Mean Square Error (MSE) between the original image/signal and the reconstructed one. As can be seen, compared to SNR/SDR metrics, PSNR focuses more on the peak error between the original and reconstructed signals, providing a measure of the maximum possible error.

$$\text{PSNR} = 10 \cdot \log_{10}\left(\frac{\text{Max}^2}{\text{MSE}^2}\right), \quad \text{MSE} = \frac{1}{mn}\sum_{i=0}^{m-1}\sum_{j=0}^{n-1}[I(i, j) - K(i, j)]^2. \tag{2.15}$$

Structural Similarity Index (SSIM): As defined in Eq. 2.16, this metric is used to measure the similarity between two images. It is designed to improve traditional methods like MSE and PSNR by considering changes in structural information, luminance, and contrast. In this formula, μ_x and μ_y are the mean values of images x and y, respectively. σ_x^2 and σ_y^2 are the variances of x and y, respectively. σ_{xy} is the covariance of x and y. Finally, C_1 and C_2 are constants to stabilize the division with weak denominator, typically defined as $C_1 = (K_1 L)^2$ and $C_2 = (K_2 L)^2$, where L is the dynamic range of the pixel values (e.g., 255 for 8-bit grayscale images), and K_1 and K_2 are small constants (e.g., $K_1 = 0.01$ and $K_2 = 0.03$).

$$\mathrm{SSIM}(x, y) = \frac{(2\mu_x\mu_y + C_1)(2\sigma_{xy} + C_2)}{(\mu_x^2 + \mu_y^2 + C_1)(\sigma_x^2 + \sigma_y^2 + C_2)}. \tag{2.16}$$

Standard Deviation Error: As defined in Eq. 2.17, this metric refers to the *Standard Error of the Mean* (SEM), which is a measure of how much the sample mean (average) of a dataset is expected to vary from the true population mean. In this formula, x_i is each individual value. $\bar{x}$ is the sample mean, calculated as $\bar{x} = \frac{1}{n}\sum_{i=1}^{n} x_i$. Finally, n is the number of values in the sample. It estimates the uncertainty associated with the sample mean due to random sampling error. This metric has been used to measure the error of the Approximate Entropy application as a case study of our proposed approximate designs in Sects. 5.2 and 5.3.

$$\text{Standard Deviation Error} = \frac{\sigma}{\sqrt{n}}, \quad \sigma = \sqrt{\frac{1}{n-1}\sum_{i=1}^{n}(x_i - \bar{x})^2}. \tag{2.17}$$

Root Mean Square Error (RMSE): As defined in Eq. 2.18, this metric is a commonly used measure of the differences between values predicted by a model and the values observed. It is the square root of the average squared differences between prediction and actual observation. RMSE is a good measure of how accurately the model predicts the response and is the standard deviation of the prediction errors (residuals). In this formula, x_i is the actual observed value. $\hat{x}_i$ is the predicted value. n is the number of observations. This metric has been used to measure the error of clustering in the K-means clustering application as a case study of our proposed approximate designs in Sects. 5.2 and 5.3.

$$\mathrm{RMSE} = \sqrt{\frac{1}{n}\sum_{i=1}^{n}(x_i - \hat{x}_i)^2}. \tag{2.18}$$

2.6 Structure of Multi-kernel Applications

2.6.1 *Heartbeat QRS Detection with Pan-Tompkins Algorithm*

It has been shown that *processing stage* consumes 70% of the total energy in wearable health nodes [Bra+17]. Interestingly, both sensing and the processing algorithms in bio-signal analysis exhibit error-resiliency [PRS19b, Bas+17], making them potential candidates that can benefit from approximation. The widely-used Pan Tompkins algorithm [JJ85] detects the main heartbeat peak (QRS) in an ECG signal (cardiac arrhythmia results in incorrect ECG wave). Pan-Tompkins application, shown in Fig. 2.7, serves as the main standard of QRS detection in wearable devices. This algorithm determines the number of QRS peaks sampled in five stages: ECG signals are first passed through an 11-tap low-pass followed by a 32-tap high-pass filter to attenuate high- and low-frequency noises. The filtered stream is fed to a five-tap digital differentiator to extract the QRS slope information. Squaring is applied to amplify the signal strength and emphasize the higher frequencies, i.e., QRS points in the com-

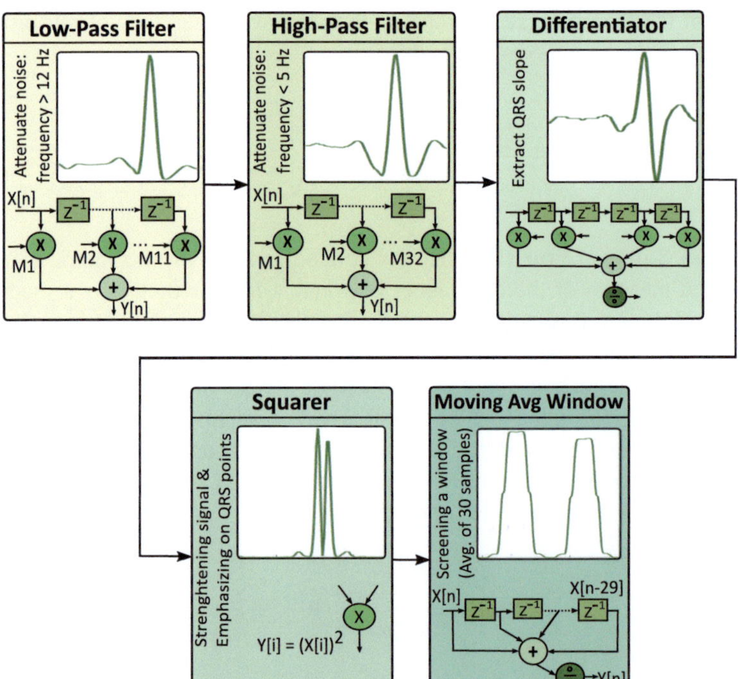

Fig. 2.7 Kernel structure of Pan Tompkins QRS detection application. All the multiplications and divisions in this application are approximated in this book

plete P-QRS-T complex [PRS19b]. Finally, R peaks are detected by screening the signal (averaging samples to the width of the moving window).

2.6.2 *Joint Photographic Experts Group (JPEG) Image Compression*

This compression technique is used in many image/video processing programs. JPEG algorithm, shown in Fig. 2.8, is initiated by RGB to YCbCr color conversion which converts the image colors from the RGB (Red, Green, Blue) color space to the YCrCb color space. In the latter, the luminance component (Y) reveals the brightness information, while the two chrominance components (Cr and Cb) represent the red- and blue differences, respectively. Such conversion makes it easier to perform operations such as compression and clustering, especially when color information needs to be processed independently of intensity. This is followed by applying 8×8 2D-DCT to identify and remove insignificant portions of information in the image without a significant loss in the visual effect. The main compression takes place via quantization, in which each DCT-converted data is divided into the user-defined compression ratio (a.k.a quality factor). In this book, the normal compression ratio of 20:1 is adopted, which lies in the range of acceptable compression rates for aerial applications [KGa+19]. Afterward, the quantized coefficients are sorted in a zigzag sequence pattern with a frequency-ascending order (low-frequency components, which are more important to the human visual system, will be retained through this re-arrangement, and others will be discarded). Finally, the Huffman scheme is applied for encoding the image.

2.6.3 *Harris Corner Detection (HCD)*

Harris Corner Detection (HCD) is ubiquitously utilized in tracking algorithms and as a feature extraction algorithm in many vision-based programs such as indoor localization and object tracking in Uunmanned Aerial Vehicle (UAV) due to its robustness

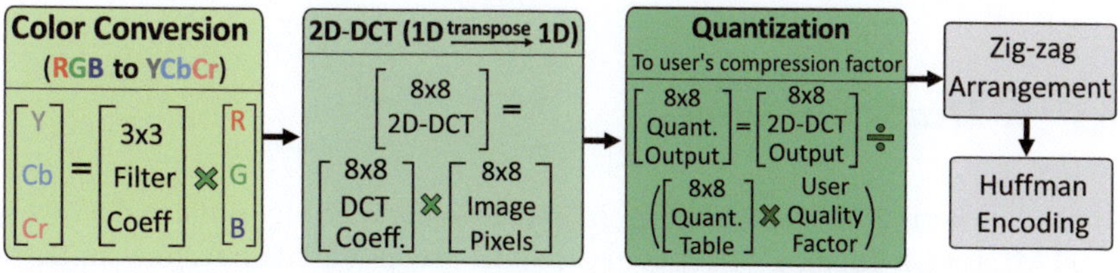

Fig. 2.8 The structure of JPEG Compression application. All the multiplications and divisions in this application are approximated in this book[Ebr+21]

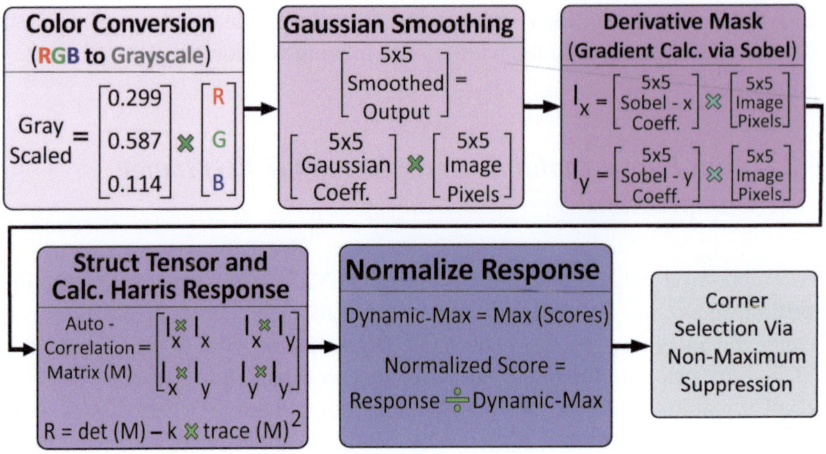

Fig. 2.9 The structure of Harris Corner Detection application. All the multiplications and divisions in this application are approximated in this book [Ebr+21]

in detecting corners in noisy images captured from on-board camera [HS88]. This technique locates the corners, i.e., pixels having strong intensity variations with their local neighborhood, in both vertical and horizontal directions. The extracted corners are then joined in consecutive frames, forming a vector in the direction of movement. As illustrated in Fig. 2.9, this algorithm includes six steps: The image is first converted from RGB to grayscale to reduce the load of computations. Then, a 3×3 Gaussian filter is applied to remove the noise of images taken from the onboard camera. This pre-processing step is reported to remarkably improve the quality of corner detection [SMS18]. Afterwards, the gradient of the image in both horizontal and vertical directions is computed using Sobel operators. These derivatives are then multiplied by each other to construct the so-called auto-correlation matrix. Then the corner strength (a.k.a Harris response) is calculated, which reveals the intensity of changes to surrounding pixels: Negative response value means the pixel probably belongs to an edge, while small and large positive, empirically chosen, values imply that the pixel belongs to a flat region and corner, respectively. Finally, the corners represented by the highest-value responses are determined via Non-Maximum Suppression, which finds the maxima in windows of 5×5 pixels.

2.6.4 K-Means Clustering

K-means clustering (illustrated in Fig. 2.10) is a widely used algorithm in data analysis and ML for partitioning a dataset into distinct groups based on feature similarity. This algorithm is used in many domains, including image processing (e.g., medical imaging and object recognition), security (fraud detection, anomaly detection),

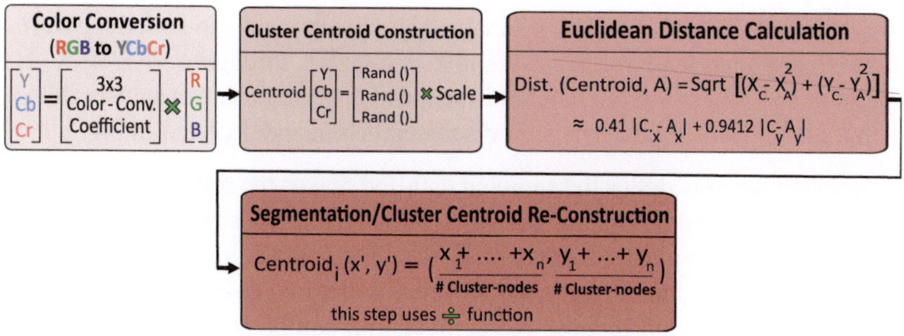

Fig. 2.10 Kernel structure of K-means application. All the multiplications and divisions in this application are approximated in this book

pattern recognition (speech, handwritten), data reduction, etc. It has four kernels as follows. Color Conversion to YCrCb (which is explained earlier). Cluster Centroid Construction: Determines the initial positions of the cluster centroids (i.e., each is the central point of a cluster, representing the mean position of all the clusters' points). A random initialization, therefore, provides a starting point for the iterative K-means clustering process. The centroids represent the central points around which clusters will be formed. Euclidean Distance Calculation: Measures distances between data points and centroids to assign points to clusters. In fact, this step determines which cluster each data point belongs to by finding the nearest centroid. This step is crucial for assigning data points to clusters. Segmentation: This process divides an image into segments or clusters based on the color information or other features. Then, each pixel or data point is assigned to the cluster whose centroid is closest based on the distance calculation. Therefore, segmentation groups similar data points together based on their feature values, leading to the formation of clusters. After all data points are assigned to clusters, the centroid of each cluster is recalculated as the mean of all data points assigned to that cluster. This process is called Cluster Centroid Reconstruction, which updates centroids based on current cluster assignments to refine the clustering result.

2.6.5 *Blind Source Separation (BSS)*

Consider n sources of signals sampled in m time slots; the data matrix size is $S_{n \times m}$. Due to the mutual interference of original signal(s) and possible noises emitted by other sources, the sa data, is inevitably a mixture of multiple sources (mixed-data matrix X). The relation between S and X can be formulated as $X = A \times S$, wherein A is the interference matrix that distorted the original signal. In reality, the coefficients of the original signal S and the interference matrix A are unknown, making the receiver *blind* to the source signal. BSS attempts to invert the mixing process and recover a near-optimal estimation of the original signal S. Mathematically this can be formulated as $\widetilde{S} = A^{-1} \times X$. This is obtained by calculating the so-called *separation-matrix* W and then applying it on the mixed-signal X through $\widetilde{S} = W \times X$. BSS consists of two stages—whitening and Independent Component Analysis (ICA)—which together form eight kernels, as discussed herein and depicted in Fig. 2.11.

Whitening process decorates the sampled data and makes the sources as statistically independent as possible. Whitening is applied through the five steps. First, *covariance matrix calculation* captures the statistical relationships, i.e., deriving a covariance C between different data features. After transforming the covariance matrix to the *Hessenberg* form (which reduces the computation complexity of the next phase), the *QR eigenvalue and eigenvector decomposition* steps break down the covariance matrix C into its constituent eigenvectors and eigenvalues ($C = E \times D \times E^T$, where E is the matrix of eigenvectors and D is the diagonal matrix of eigenvalues). The set of eigenvectors represents the directions along which the data varies the most, while the corresponding eigenvalues determine the variance along each of those eigenvector directions. Finally, *Back Substitution Solver* derives the *whitening* matrix W as ($W = D^{-\frac{1}{2}} \times E^T$) and apply it to the zero-centered input data to decorrelate the data and remove the dependencies between them.

After whitening, ICA process should be applied. FastICA is the most widely used approach for solving this extraction problem. FastICA algorithm iteratively enhances a randomly-initiated matrix until convergence happens. Each iteration has three sub-steps, as follows. *Tanh Analysis* is an optimization step that improves the algorithm's convergence by maximizing the data's statistical independence (or non-Gaussianity) and reducing the separation matrix's sensitivity to outliers. Afterward, *Evolving Separation* refines the estimation of the independent components by updating the separating matrix (i.e., applying a contrast function on the current separation matrix). After performing the update process, the estimated independent components may still be correlated. Therefore, the *Gram-Schmidt decorrelation* algorithm [Bjö94] is applied to ensure that the components are decorrelated, i.e., orthogonal to each other. Finally, the independent components are normalized, the weight vectors in the separation matrix are updated, and the convergence condition is assessed (by comparing the Euclidean distance of two consecutive separation matrices).

2.6.6 Image Classification via CNN

Image classification is a ubiquitously-used image processing application that is also highly used for evaluating the efficacy of various approximation techniques. In this context, a lightweight CNN model called MNIST-CNN [CT16] is adopted, which basically is a Multi-Layer Perception (MLP) implemented in Python for the classification of the Modified National Institute of Standards and Technology (MNIST) Digit dataset [LDe12]. Figure 2.12 shows that the MLP consists of two fully connected hidden layers with 100 and 32 nodes, respectively. The input and output layers

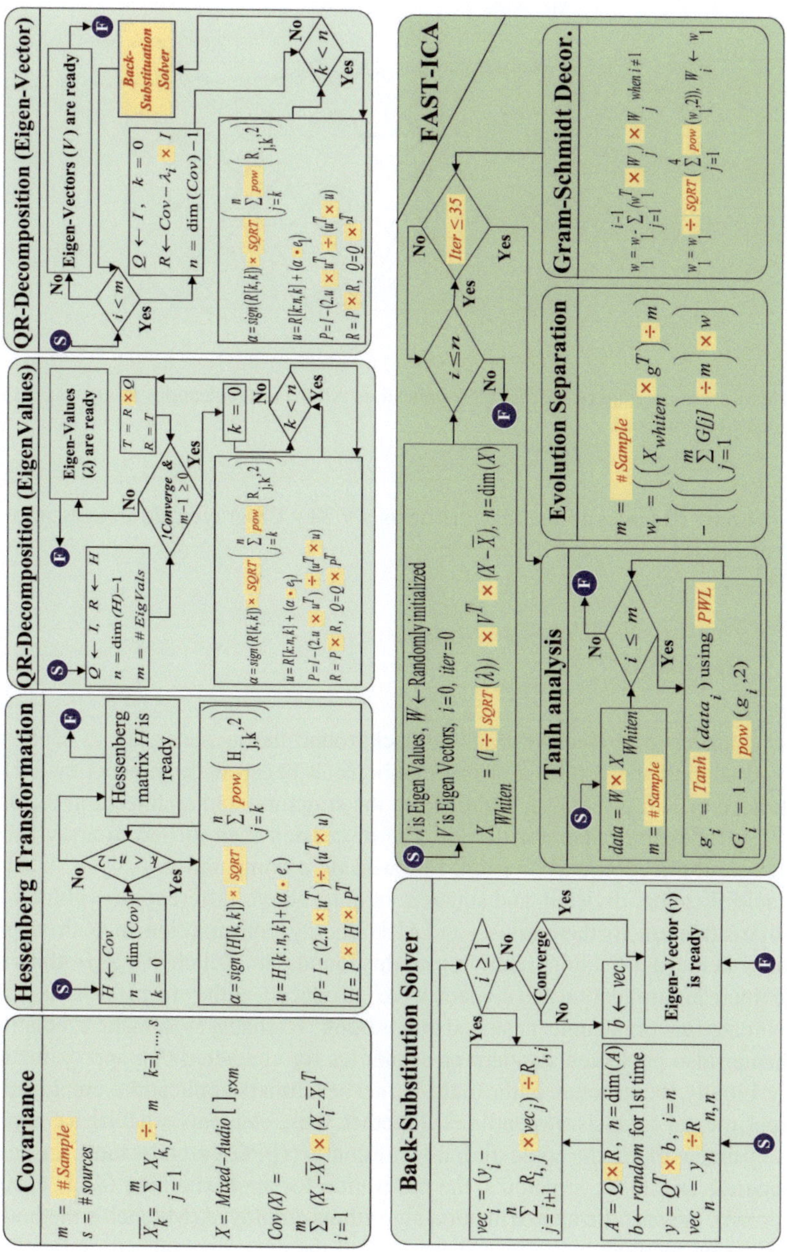

Fig. 2.11 The kernel structure of Blind Source Separation application (whitening and FastICA) and the detailed instruction flow in each kernel. The approximation knobs (utilized in Chap. 7) are highlighted in the figure

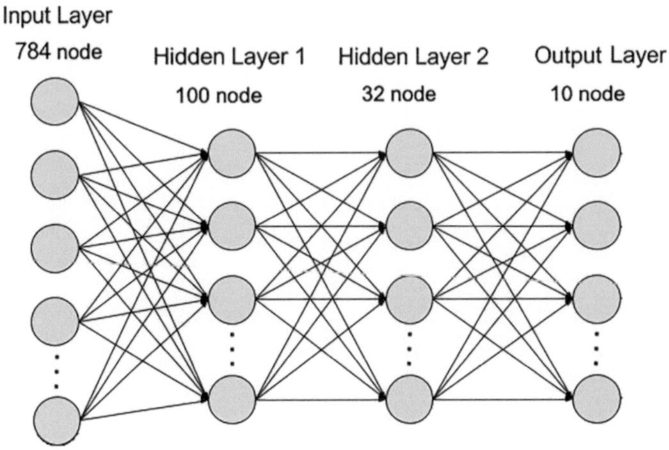

Fig. 2.12 Kernel structure of MNIST NN application. All the multiplications inside neurons are approximated in this book

have 784 and 10 nodes to classify the dataset's 28×28 gray-scale images into 10 different classes.

2.7 Conclusion

This chapter has provided the essential background that is necessary to contextualize and support the contributions of this book. It began with an overview of the architecture of Xilinx FPGAs, highlighting the structural and architectural features relevant to arithmetic operator design. The discussion then moved to array-based accurate multiplication and division, followed by a comparative analysis of addition, multiplication, division, and square root units across different bit-widths, with particular attention to their impact on ALU energy consumption in both single- and multi-kernel applications. Building on this foundation, Mitchell's algorithms for approximate multiplication and division were introduced, as they form the basis of the approximate functional units designed in this book. To enable systematic evaluation, the chapter also presented standard error metrics for characterizing approximation quality. Finally, the structure of the multi-kernel benchmark applications employed in the experimental analysis was outlined. Together, these elements establish a comprehensive framework for the book: the hardware context (FPGA architecture), the baseline accurate arithmetic structures, the motivation for approximation (ALU energy breakdown), the specific approximation algorithms employed (Mitchell's methods), the evaluation methodology (error metrics), and the experimental setting (application kernels). This foundation paves the way for the subsequent chapters, where the proposed approximate designs and their evaluations are presented in detail.

Chapter 3
Related Work

Abstract This chapter surveys prior work on approximate computing techniques that enable low-latency, high-throughput, and energy-efficient accelerators for error-resilient applications. It situates the book's cross-layer methodology and reconfigurable architectures within a broad landscape of approximation strategies spanning data acquisition, hardware, and software processing. The discussion covers application domains such as image and bio-signal processing, audio, machine learning, biometric security, big data analytics, networking protocols, and reliable mission-critical systems, highlighting how approximation has been tailored to each. Techniques are grouped by abstraction level into circuit-, architecture-, and application-level methods, as well as cross-layer approaches that combine them. The chapter emphasizes that effective cross-layer design requires sensitivity analysis to tune approximation knobs and design-space exploration heuristics to efficiently identify Pareto or near-Pareto configurations that balance performance gains against accuracy loss. It concludes by outlining the specific families of techniques and in-network computing-oriented methods that are most relevant to the contributions developed in the remainder of the book.

Keywords Cross-layer approximation · In-network computing · Distributed computing · FPGA/CGRA · SIMD/MIMD · High-throughput · Pipeline · Mitchell's logarithmic multiplier-divider · Error sensitivity analysis · Multi-objective heuristic · Design space exploration · Neural networks · Classification · Programmable networks · P4

3.1 Introduction

Chapter 1 has introduced the approximate computing paradigm as a viable solution to design low-latency, high-throughput, resource-, and energy-efficient hardware accelerators for error-resilient applications. This book contributes in the same direction by presenting a cross-layer approximation methodology based on novel reconfigurable circuits and architectures. The proposed architectures and methodology are also evaluated on various signal processing applications.

Z. Ebrahimi and A. Kumar, *Cross-Layer Approximation and In-Network Acceleration*,
https://doi.org/10.1007/978-3-032-21712-7_3

This chapter provides a summary of previous work related to the contributions presented in this book. In general, the approximation techniques have been applied in diverse domains of applications. Few exemplar domains are image processing [LL22, HJi+20], bio-signal processing [EK24, EK21, PRS19b], audio processing [Liu+19a], ML models [Liu+21, DHP21], bio-metric security (e.g., iris scanning) [Has+18], big data [MH21], networking and communication protocols [Ven+15a, Qad+17, Joh+20, AC19], and RMS applications [XMK16].

As shown in Fig. 3.1, the approximation techniques in the literature cover various execution phases, from data acquisition to hardware and/or software processing.[1] Approximate sampling and compression are the techniques that are applied in the data acquisition phase (discussed in Sect. 3.2). Techniques for approximate *processing*—that are applied on different levels of computing stack—can be classified in circuit-, architecture-, or application-level (discussed in Sects. 3.3–3.5). Some works have also targeted the combined utilization of some of these techniques across the abstraction layers, which is referred to as cross-layer approximation [MSh+16] (Sect. 3.6). The pre-requisites for such a cross-layer design approach are (1) a sensitivity analysis scheme to appropriately adjust the approximation knobs (which is discussed in Sect. 3.7) and (2) a DSE heuristic (to prevent an exhaustive DSE) which can find the Pareto or near-Pareto approximate configurations that maximize the performance-gain while minimizing the accuracy loss (Sect. 3.8). The following provides a brief

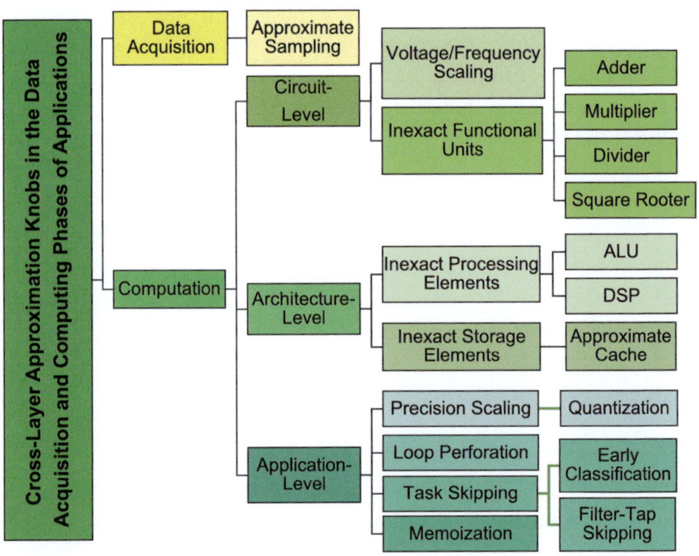

Fig. 3.1 Classification of the commonly-used approximation techniques at different abstraction layers in the sensing and computing phases of applications

[1] Detailed surveys on these techniques can be found in [Mas+24, Dam+24, Que+23, Dam+23, Leo+23, Arm+22, Zer+21, Sek21, RLB20, HJi+20].

description of the widely-used techniques in each class that are related to this book, along with those targeted for INC (Sect. 3.9).

3.2 Approximate Data Acquisition

Modern computing systems—from high-performance servers to embedded processors—utilize 32-bit and 64-bit operations. Such systems utilize various number representations such as IEEE single- and double-precision floating-point and 32-bit and 64-bit integer schemes for performing arithmetic operations. However, such high-precision number schemes are usually not required for most error-resilient applications, such as image, audio, bio-signal processing, and machine-learning models. In fact, many of such data, e.g., bio-signal, often contain substantially correlated or redundant data, partly due to their periodic structures.

Furthermore, the high complexity and memory storage requirement for performing the computations at such high precision prevents deploying them to the resource-constrained embedded systems and network switches. In fact, some studies have demonstrated that the data acquisition stage dissipates significant energy in the end-to-end energy footprint of the systems, e.g., about 30% of total energy in wearable gadgets employed for bio-signal processing [Bra+17]. In this context, to reduce the memory and computation requirement, the designers are forced to use integer or fixed-point precision [Pat+22, XZ19] (by converting the analog signals around us, such as audio and bio-signals, to the digital domain using an ADC), which are a realization of precision scaling techniques which is a widely explored approximation techniques and further discussed in Sect. 3.5. Finally, reducing the number of samples can significantly improve the processing requirement and response time of these applications [RBD13].

Approximate sampling strategies can be classified into two groups: *Random* and *non-random*. The first category selects data points randomly, without considering the signal structure or the significance of the data. An example of this category is Monte Carlo sampling used in numerical simulations. The non-random category can be further divided into *stratified* and *adaptive* approaches [Dal+24]. *Stratified* methods divide the signal/dataset into stratum or segments and then pick multiple samples from each group. This method ensures equal representation of data from different groups. *Adaptive* methods dynamically adjust the sampling strategy based on selective data characteristics. Among these, compressed sampling is the most commonly used technique, as it samples only the most essential parts of the signal and attempts to reconstruct the missing parts mathematically. Compressed sensing itself can be further divided into lossless and lossy approaches. The former ensures that no data is lost, enabling full signal reconstruction by randomly selecting a subset while preserving statistical properties. In contrast, the latter discards some data permanently to improve efficiency, often by dropping less significant data points to save memory. The combination of random and non-random sampling methods has also been explored by researchers. For example, StreamApprox [Quo+17] proposed an adaptive reservoir sampling algorithm that performs the computation over disjoint sub-streams from various sources. In Chap. 7 of this book, a *uniform* stratified approach is adopted for reducing the number of audio samples in an INC-setting. The reason for choosing the uniform method is to maintain a homogeneous proportionality of samples from the whole dataset.

3.3 Circuit-Level Approximation Techniques

Detailed reviews that have specifically focused on circuit-level imprecise designs can be found in [Jia+17, Che+22a]. For the quantitative evaluation of some of these designs, readers may also refer to [HJi+20]. Researchers have extensively explored the circuit-level approximation track throughout the last decades. The studies can be divided into three sub-classes, i.e., DVFS, circuit pruning, and design of inexact operations.

3.3.1 Dynamic Voltage and Frequency Scaling (DVFS)

This technique falls into the category of non-deterministic approximation and is used as a power management scheme that dynamically aligns the energy consumption of a circuit with the desired performance [NT17, HM07]. Consider the dynamic power consumption of a circuit, demonstrated in Eq. 3.1, where V_d is the supply voltage, C is the circuit capacitance, and f is the circuit operational frequency.

$$P_d \propto C V_d^2 f \tag{3.1}$$

As can be observed, due to the quadratic relation between V_d and P_d, the dynamic power consumption of a circuit can be significantly alleviated by reducing its supply voltage, a.k.a voltage over-scaling or under-volting. Some works have exploited this technique to reduce the energy consumption of the error-resilience parts of the applications. For example, the authors of [Chi+14] exploit under-volting for their proposed recognition and mining processor, specifically, for approximate MAC units. However, there are shortcomings associated with such techniques, as discussed below. First, applying the under-volting scheme also increases the transition time, i.e., the charging and discharging time of the logic gates, which impacts the maximum allowed operating frequency of the circuit. Second, implementing effective DVFS strategy requires sophisticated algorithms to balance the power saving while satisfying the performance requirement of the system. Finding the optimal voltage and frequency settings for varying workloads is a tedious job. Second, the effect of DVFS is diminished with the technology advancement. In fact, as the transistor feature size reduces, the share of static power (which is marginally affected by DVFS) goes above the dynamic power. Third, under-volting can result in non-deterministic timing violations, for which additional circuitry is required to control/recover for the resulting accuracy drops [Sch+17, Sta+20]. Finally, DVFS may not yield significant energy savings in systems where static (leakage) power dominates dynamic power consumption or when applied to light workloads. In such cases, the overhead of frequent voltage/frequency transitions can limit potential gains, and alternative strategies like power gating may be more effective.

3.3.2 Inexact Functional Units

The methods in this category involve pruning (i.e., removing) or simplifying parts of the circuit that significantly impact performance metrics (such as resource footprint, power dissipation, and critical path delay) while minimally affecting the ultimate output accuracy [Ven+15b]. Pruning techniques have been mostly applied at the transistor- or gate level to widely used arithmetic operations such as addition and multiplication. While removing transistors has mostly been adopted in a way to reduce the critical path of the target small-scale circuit (e.g., in an accurate FAs [Gup+13, Yan+13]), gate-level pruning has also been adopted to remove the gates having the highest contribution in e.g., dynamic power dissipation. For example, the work of [Sch+17] has proposed a framework to analyze the generated netlist from the accurate implementation of an operator and rank the different gates based on their switching activity (toggle count). The framework then determines the significance of the gates as follows: All the gates that contribute to the computation of output bit O_{i+1} have a significance twice higher than those that contribute to the generation of output bit O_i. Afterward, the gates with a lower ranking are pruned, and the resulting approximate circuit is assessed in an accelerator designed for DCT. The main shortcoming of the pruning approach at the transistor- or gate-level is its lack of scalability to more complex architectures. In the following, the pruning and simplifying approaches that are targeted for each category of functional units are described.

Approximate Adders The studies in this track have mainly adopted three approximation approaches [DNT16, DNT17]. (1) To reduce the overall critical path delay of the adder, many studies have broken down the carry propagation path of the adder into segments or even completely eliminated the carry chain. (2) In order to enhance area-efficiency, the Karnaugh map of FAs and/or HAs have been simplified. Such FAs/HAs are then utilized for the addition of LSBs in a higher-order adder. (3) Finally, configurable-accuracy adders have been designed by augmenting inexact sub-adders, which are augmented with error detection and correction modules. For example, the approximate GeAr design [Sha+15] implements a larger adder via utilizing multiple sub-adders in parallel, with no carry propagation between the sub-modules. In this design, all sub-adders (except for the least significant adder) utilize a number of previous bits to predict the missing carry at design time. Furthermore, some of these works, including GeAr, also presented extra circuitries to generate accurate output when needed. However, such error correction modules come at the cost of degrading the performance of the approximate design. Following the same elimination scheme, the works of [Bec+16, Pra+18, Ull+18, SSA18] have proposed various LUT-based approximate adders for FPGAs, by directly configuring the 6-LUTs (Table 3.1).

Approximate Multipliers As discussed in Chap. 1, multipliers and dividers are the most energy-hungry units that extensively employed in the main kernels of a wide domain of error-resilient programs.[2] This carves out a prominent niche for their approximation. For example, Deep Residual Learning (ResNet152) [He+16] has 11.3 billion MAC operations per forward pass for the processing of a single image. Therefore, approximate versions of the multiplier and divider can be utilized to obtain area-optimized, low-latency, and energy-efficient implementations. Previous works have proposed different approximate multiplier and divider designs. A detailed survey on the design approaches for imprecise multipliers can be found in [Wu+24]. However, as discussed in Chap. 3, the approximation techniques presented in most of these works focus on ASIC-based systems and ignore the architectural specifications of FPGAs or render limited performance gain. Therefore, the techniques designed for ASICs are less effective in gaining similar gains when directly deployed and used in FPGA-based systems.

PP approximation: This class of works has applied approximation on the generation, reduction, or addition of the PPs. For the first category, techniques such as truth table simplification in 2×2 and 4×4 multipliers are employed. These units are then used in a hierarchical approach to build higher-order multipliers [SRe+16, Liu+14, KGE11, VMr+17]. For the *accumulation* or *reduction* of PP rows (which is the most computational-intensive process in array-based multiplication approaches), 3:2 and 4:2 imprecise compressors [MSt+20, OAk+17, SS17, DEs+18, MWa+20, JSD20, VL20] are designed and exploited in array-based multipliers. The approximate 4:2 compressor is one of the most explored techniques in this regard. An accurate 4:2 compressor generates a 2-bit output (sum and carry) and an output C_{out}. All three bits will be used only if all input bits have the value of 1, and these three bits will be equal to 100_2. Therefore, to apply approximation, the imprecise 4:2 design proposed in [LL13] truncates this value to two bits, i.e., 10_2. Some following works [SAn+18, SS17] have also reduced the error of the approximate 4:2 compressors by computing the propagate and generate signals from the generated PPs. Finally, for the *addition* of the generated PP rows, breaking the carry propagation path is adopted by some studies [YHS20, Ull+18]. The main shortcoming attributed to these classes of works is their limited scalability when transported to a larger hierarchical design. For example, the simplification of Karnaugh map or PP tree should start from scratch for each size of multiplication, otherwise error magnitude may drastically increase as it becomes accumulated in a recursive design approach. Moreover, compressor-based designs are either usually heavy on the ARE

---

[2] Section 2.3 compares the area, delay, and energy of addition, multiplication, division, and square root operations in both ASIC and FPGA for different bit-width precision along with the cumulative energy of these operations in various single- and multi-kernel applications.

Table 3.1 Summary of approximation approaches for ASIC- and FPGA-based multipliers and dividers

Approach	MUL/DIV	P[a]	S[b]	ARE[c] up to (%)	Description	A/F[d]	Reported performance gain[e]	E[f]
Partial product gen./add./accumul.	✓/✗	✗	✗	7.6	Inexact 4:2 compressor in Dadda MUL [MSt+20, JSD20, DEs+18]	A	{A, D, P} +	✗
				1.7	Utilizing inexact compressor in 3/4 of LSB columns [MWa+20]		{A, P} ++, D +	✗
				8.4	Inexact 4:2 compressor in Booth MUL [SAn+18]		A +, E ++	
				Conf.	Library of inexact MULs based on 2 × 2 instances [VMr+17, SRe+16]		{A, D, P} ++	
				0.3	Cutting carry propagation in 4/8-bit array MUL [YHS20, Ull+18]	F	{A, D, P} +	
				8.5	Inexact 3:2/4:2 compressor for Dadda MUL [VL20]		{A, P} +	
				Conf.	Library of 4 × 4/8 × 8 MUL with approximate PPs [SSA18]		{A, P, D} +	
Trunc. MUL/DIV	✓/✗	✗	✗	10.9	LOD-based MUL [FFr+20], with rounding [SVa+19, RZe+17]	A	{A, P} ++	✗
	✗/✓			6.7	LOD-based DIV [HJi+19, HJi+18]		{A, D} +, E ++	
	✓/✗		✓	1-4.7	Var-prec. MUL using 8-bit truncated instances [HJi+19, HJi+18]		{A, E} +	
MUL-based DIV	✗/✓	✗	✗	2.9	Piecewise linear approx. & rounding reciprocal of divisor [Vae+18]	A	{D, E} ++	✗
				6.4	Truncating reciprocal of divisor [SVa+17]		{A, D, E} +++	
				16.3	Approximating reciprocal of divisor using truth-table [RZe+16]		{A, D} +, E ++	
		✓		4.9	Approximation of reciprocal via Taylor series [Mel+19]		{D, E} +++	✗
Mitchell's MUL/DIV Algorithms [NMi62]	✓/✗	✗	✗	2.9	Enhance accuracy of Log in piecewise approximation [SFJ19]	A	{A, P} ++, D +	✗
				> 3.9	Use different approximate Adds in Mitchell's MUL [WLi+18]		{A, P} +++	
				2.7	Add a error-correction term to Mitchell's MUL [HHS18a]		{A, P} ++	

(continued)

Table 3.1 (continued)

Approach	MUL/DIV	P[a]	S[b]	ARE[c] up to (%)	Description	A/F[d]	Reported performance gain[e]	E[f]
				2.7	Adding up to 256 error-coefficient to Mitchell's MUL [HSa+20]		{A, P} +	
	✗/ ✓			3.0	Add one error-correction (similar approach to [HHS18a]) [HHS19]		A +, {D, E} ++	
	✓/ ✗	✗	✗	1-1.6	Adding one to five error-coefficient to Mitchell's MUL [ZSA20]	F	{A, E} ++, T +	✗
9BF29E	✓/ ✓	✗	✓	**0.8**	**64 error-coeff. for MUL/DIV [Ebr+21, EUK20, ZSA20]**	**F/A**	**{A, E, T} +++**	✓
9BF29E	✓/✓	✓	✗	**0.6-1**	**Pipelined MUL/DIV with tunable accuracy [Ebr+22]**	**F**	**{A, E, T} +++**	✓

[a] Pipelined, [b] SIMD, [c] Average of absolute relative error (a.k.a MRED), [d] ASIC/FPGA, [e] Area/delay/energy/power/throughput, [f] End-to-end performance gain evaluation

index or render moderate performance gains, especially when only applied on a few least significant PP columns. Some of these works have also provided libraries of approximate multipliers with various resource-accuracy trade-offs.

Modular design based on small-scale inexact multipliers: To implement higher-order multipliers the works of [BMH14, KGE11, SRe+16] have adopted a modular design approach, wherein a 2M×2M multiplier is implemented via four M×M instances, as shown in Fig. 3.2. For example, the authors of [KGE11] have proposed an approximate 2×2 multiplier, which produces a 3-bit output instead of 4-bit by approximating the accurate product value of 9 (multiplying 3 with 3) with 7. Modular approaches usually utilize an accurate sub-multiplier for the most significant one, i.e., $A_H \times B_H$ part in Fig. 3.2 (to prevent significant accuracy drop degradation in the output). At the same time, the remaining non-significant sub-multipliers are substituted with inexact versions. The performance gains from these approaches are moderate due to limiting the approximation to only a part of the design.

Resizing to smaller multiplier: Another scope of work has explored resizing the larger multipliers to smaller ones. These studies separate the input operands into significant and non-significant parts. A smaller (usually accurate) multiplier is then utilized to multiply the significant parts (i.e., the most significant bits). Finding the most significant part of each operand has been done statically (a user-defined bit position [Nar+15, KGY10]) or dynamically (based on the position of the leading one [SVa+19, FFr+20, RZe+17, SIS15]). For the implementation of Leading One Detector (LOD) and extracting fractional parts, the ASIC works have utilized large priority encoders and barrel-shifters [LB12]. However, the ASIC-based approaches weakly fit FPGAs due to using layers of multiplexers (poorly mapped on LUTs [Moc+12]). Targeting FPGAs, authors in [Li+08] proposed an 8-bit LOD which uses consecutive levels of LUTs to implement each bit of LOD-output since each bit is a function of all bits in the input. The negative result of this cascading is the exacerbation of circuit latency. Moreover, using large barrel shifters for re-aligning fractional parts is inevitable, which further increases overhead. In another track, studies such as [Cho+04, SVK07, WKL11, JTT03] have truncated the computation of the least significant product bits. Some of these works have also proposed error compensation schemes to reduce truncation-induced errors in the final product. Although offering high resource savings, these truncation techniques suffer from errors up to 100%.

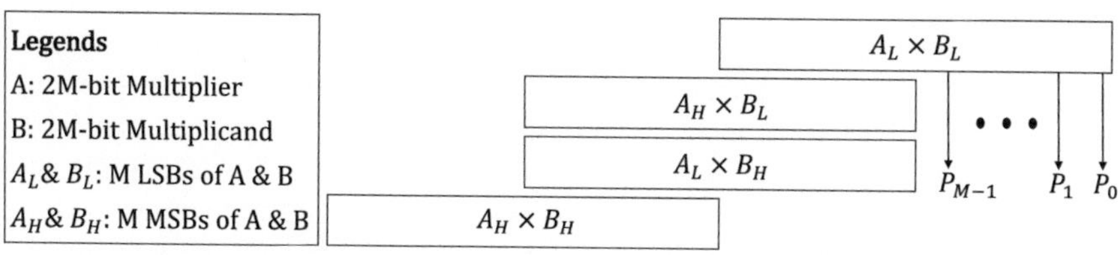

Fig. 3.2 Designing higher order multipliers from lower order multipliers

Logarithmic multiplier: In a more recent trend of studies, some works have adopted the approximate Mitchell's algorithms [NMi62], which translate multiplication to addition in the logarithmic representation. This algorithm is explained in Sect. 2.4 and briefly formulated here:

$$P = A \times B \xrightarrow[\text{Log}]{\text{Approx}} .\widetilde{\text{Log}_P} = \widetilde{\text{Log}_A} + \widetilde{\text{Log}_B} \tag{3.2}$$

Such a transformation generally offers higher performance gains, by reducing it to that of a same-sized multiplier. Such improvements, albeit, come at the cost of relatively higher error than some array-based approaches (the AREs of Mitchell's multiplication is 3.8%, respectively). In this context, various schemes have been recently proposed to reduce this error, which is assessed in both ASIC- and FPGA-based implementations.

As discussed in Sect. 2.4, Mitchell's multiplication (division) algorithm consists of four steps: Finding the leading one as the integer part, re-aligning the rest of the bits as the fractional part, the addition (subtraction) of both parts for the two operands, and finally shifting the summation (subtraction) of fractional part w.r.t the summation (subtraction) of integer parts. Works like [NMi62] investigated piece-wise linear error-reduction schemes to calculate log and anti-log individually for each segment within power-of-two-interval. These methods are applied to each of multiplier's input separately, *neglecting magnitude of error after multiplication or division*. Tackling piece-wise approximation overhead, the MBM ASIC multiplier [HHS18a] has proposed a single error-correction term (for all input combinations), which is added once after computing summation of the fractional part. However, as a single error-reduction term weakly fits all input combinations, this approach has eventuated in many output overflow cases when added to the original designs. To alleviate such a problem and also target FPGAs, various FPGA-specialized designs, LeAp [ZSA20], SIMDive [EUK20], and RAPID [Ebr+22] are proposed, which are detailed in Chap. 4.

FPGA-customized multipliers: As mentioned in Chap. 1, the direct deployment of ASIC-based approximate approaches into FPGAs does not yield the same performance improvements as in ASIC implementations [Ull+18, SSA18]. This is primarily due to the underlying architectural differences between these platforms' hardware. To cope with this challenge, some recent works have narrowed their focus toward designing circuit-level approximation approaches, specifically for FPGA, by taking into account the LUT-based structure of FPGAs. These works have mainly focused on array-based or Booth-based multiplication approaches. For example, some works have truncated the carry propagation and the LUTs utilized for implementing the LSBs of the product in 4×4 multipliers (these designs have been customized for FPGAs for array-based [YHS20, Ull+18, SSA18, Ull+21] and Booth-based [SUl+20] architectures). Furthermore, inaccurate 3:2 and 4:2 compressors have been recently proposed in [VL20], geared toward an LUT-oriented implementation. Despite the specialized customization in these schemes for reconfigurable platforms, their resource savings have not been significant (when targeting a low error-index).

Moreover, Booth-based designs suffer from a long critical path [SUl+20]. Finally, none of these multipliers are designed to address the demand for high-throughput computing, whether through pipelining or SIMD parallelization. Such drawbacks have therefore accentuated the demand for exploring other techniques which can enable higher savings with an acceptable accuracy. Therefore, in this book, such call are attempted to be addressed by designing LUT-based logarithmic multipliers (which are also further customized for SIMD execution approach and pipelining).

Approximate Dividers and Square Rooters The accurate division **Division with inexact subtractor**: This family of dividers has replaced the precise subtractors with imprecise counterparts in the 2N-by-N non-restoring and restoring array dividers [ESS20, SES19, LCh+18, LCh+16]. Generally, the array-based approximate dividers offer high accuracy; however, the resource savings delivered by them are not significant due to maintaining the array structure [HJi+19, HJi+18]. Furthermore, they do not significantly improve the latency, so the delay of a divider remains multiple times larger than of its peer-sized multiplier. Finally, the overflow cases are also large in the conventional restoring array dividers (when a dividend is large and the divisor is small). In contrast, the logarithmic dividers (discussed in the following) do not overflow [WLi+20].

Multiplicative dividers: For calculating division with this approach, the reciprocal of the divisor is calculated first, then multiplied with the dividend. The inexact designs in this branch of studies have mostly applied approximation on the reciprocal of divisor, using linear piecewise approximation approaches [Vae+18] or Taylor series [Beh+19]. In another study [RZe+16], the divisor is also rounded to the nearest number in the form of $\frac{2^{k+l}}{D}$ (wherein k denoted the position of leading-one), the division operation simplified from $\frac{A}{B}$ to $A \times D$. In some studies, the operands are also truncated/rounded [SVa+17]. Overall, these approaches will impose significant resource costs in FPGAs due to requiring both the reciprocal and multiplier IPs separately. Moreover, when the truncation of the divisor goes beyond a relatively small length, e.g., 4 bits, the accuracy of its corresponding reciprocal starts to degrade significantly [Vae+18].

Resizing to smaller divider: More recent truncation-based approaches, dynamically determine the position of leading one and then utilize a narrower-width accurate instance of the divider [HJi+19, HJi+18, SIS16]. Albeit they offer notable resource improvements, these truncation techniques suffer from error cases, some are equal to 100%. In addition, the latency of such dividers still remains significantly higher than that of a same-sized multiplier. In addition, the latency of an accurate smaller divider could still be multiple times of a same-sized multiplier.

Logarithmic multiplier and divider: Similar to Mitchell-based multiplication, a division is also translated to subtraction in a logarithmic representation, as follows:

$$D = \frac{A}{B} \underset{\text{Log}}{\longrightarrow} \text{Approx.} = \widetilde{\text{Log}_A} - \widetilde{\text{Log}_B} \tag{3.3}$$

Such a transformation offers a significant reduction in the division latency of an accurate divider. However, similar to Mitchell's multiplication, the ARE of this method is 4.1% for the division. Building upon a scheme similar to the MBM multiplier [HHS18a], the same authors proposed a single error-reduction term for division in the INZeD divider [HHS19]. However, since a single error-reduction term cannot adequately accommodate all input combinations, this approach frequently leads to output overflow when integrated with the original designs. To alleviate such a problem and also target FPGAs, various FPGA-specialized designs, i.e., SIMDive [EUK20], and RAPID [Ebr+22] are proposed, which are detailed in Chap. 4.

Imprecise Square Rooters: The approximation approaches for this operation have been limited to either using a non-restoring algorithm [Adv24a] or simplifying the square root algorithm to shift and subtract [HJi+19]. In Chap. 7 of this book, the first approach (due to being more accurate while still resource-friendly) is adopted, wherein the accurate multiplication is replaced with the proposed logarithmic version of SIMDive from Sect. 4.4.

3.3.3 Pipelined Accurate and Approximate Functional Units

High throughput is highly desired for a variety of application domains, e.g., image and bio-signal processing since they are repeatedly fed with a bulk of data. Therefore, in order to boost the throughput of the designs, fine-grain pipelining has recently gained attention from researchers in both industry and academia. On the industry side, major FPGA vendors such as Xilinx have designed pipelined IPs for multiplication and division [Xil15, Xil16b]. In academia, some studies have also recently [JH19a, JH19b] applied pipelining on PP generation and accumulation in the array-based multipliers. However, all these efforts have been targeted toward an *accurate* computation. In fact, limited attention has been dedicated to approximate designs. The SAADI-EC divider [Mel+19], the modified version of the DRUM multiplier [Sca+21], and [Lyu+22] have been the only works that adopted an intra-unit pipelining approach for approximate designs. Despite the possibility of increasing throughput, this work lacks a well-established pipelining strategy, as no analysis has been performed that uniformly divides the critical path of their design over a desired number of pipeline stages. For example, the SAADI-EC-P (which is the pipelined version of SAADI-EC) divides the critical path of the non-pipelined divider into three stages, i.e., normalization, multiplication, and accumulation (for adding a new iteration of Taylor series expansion), and error compensation. Such a pipelining approach has, therefore, ignored the latency of each of these stages. As a consequence, their divider is characterized by poor performance in terms of latency (due to the high clock period after pipelining). Moreover, no analysis has been reported on the architectural implementation for such a pipelined divider.

Therefore, to surmount the rest of the foregoing challenges (especially in FPGAs), the RAPID multiplier and divider [Ebr+22] are presented in Sect. 4.3, which are the first approximate designs, efficiently geared toward fine-grained pipelining.

3.3.4 SIMD Accurate and Approximate Functional Units

To harvest parallelization potentials (e.g., matrix multiplication) in many applications, the concept of SIMD has been tailored by Intel [LGB20, MLa+19] and Xilinx [Fu+17a] to provide support for double multiplication (with a common operand) within their FPGA DSPs. A similar trend has also been pursued in academia. For example, previous studies such as [Pur+08, Lan+05, Per+06, Per+04] have presented SIMD accurate soft multipliers (8- to 32-bit), targeted for FPGAs. More recently, the work of [OR19] has designed a truncated (inaccurate) SIMD multiplier to be used in programmable devices. Such SIMD designs not only harness data-level parallelism capability but also bridge the gap between circuit- and application-level approximations by supporting dynamically reconfigurable approximation [Zer+21] through allowing runtime precision-variability in a single unit. However, such SIMD units either operate in accurate mode or are customized for an ASIC implementation. Our proposed SIMDive [EUK20], presented in Sect. 4.4, addresses these limitations through its plasticity, enabling it to switch between hybrid functionalities on the fly without requiring reconfiguration, while also supporting execution in SIMD mode.

3.4 Architecture-Level Approximation Techniques

Compared to the circuit-level , limited attention has been dedicated to architecture-level works. In fact, beside approximate caches, most of the existing works have expanded prior efforts at circuit-level to design more complex components, i.e., approximate MAC and DSPs in FPGAs , or ALUs and PEs, in CGRAs which can support a SIMD or MIMD execution model. Before discussing the approximate works, the energy-efficiency approaches for accurate CGRAs are briefly reviewed. Finally, a brief summary of approximate caches are presented.

3.4.1 Energy-Efficient CGRAs

Over the last decades, CGRAs have proved viable platforms for, e.g., health monitoring applications. In fact, domain-specific CGRAs that have targeted energy efficiency are designed by specializing their architecture for various applications. For example, RAP [BJK22] and BrainWave [Bru+20, Bru+21] are customized for the acceleration of bio-signal processing kernels by extending/modifying the architecture of Blocks

SIMD CGRA [Wij+19]. While Blocks has a homogeneous structure, RAP includes heterogeneous PEs, each supporting 24-bit fixed-point accurate multiplication, division, and CORDIC operations. ASAP [Tan+22] has enabled precision-variability for 16- and 32-bit fixed- and FP numbers only by discarding/zero-padding certain mantissa bits. Besides customized PEs, other domain-specific customizations have been adopted by some of these CGRAs. For example, in targeting bio-signal processing, the network topology has been refined (by adding diagonal links), and the size of the data memory is also specifically determined based on the sampling frequency.

3.4.2 SIMD/MIMD CGRAs

Targeting a high accuracy, computations in CGRA PEs have been carried out in 32- or 16-bit integers in both bio-signal and image processing applications [Sma18]. Nevertheless, bounding the precision to 16- or 8-bit is also reported to be satisfactory in some recent studies [HSa+20, HHS19]. In this context, HEAL-WEAR [Duc+17], i-DSPs [Duc+19], Blocks [Wij+19], and PRECISION [NVB16] are the SIMD CGRAs that have been employed for the *accurate* computations while TRANSPIRE [Pra+20] supports precision conversion among FP numbers. RaPiD is an approximate CGRA that has been manufactured and commercialized by IBM for the acceleration of AI tasks [Ven+21b, Ven+20]. RaPiD incorporates 8-way PEs and Special Function Units (SFUs) which are capable of performing MAC operations on sub-integer and FP numbers, respectively. Overall, in order to enable Instruction Level Parallelism (ILP) and Data Level Parallelism (DLP) execution models and achieve high throughput, these SIMD/MIMD architectures are characterized by their high area footprint [Int20, Xil18c], stemming from separate data-paths in each of their constituent PEs.

3.4.3 Approximate CGRAs Based on Inexact PEs

Literature studies in this track are still limited to a few techniques. The first class of works has utilized a heterogeneous combination of PEs in their CGRAs. Each of these PEs consists of an inexact adder and multiplier (some also have a divider), all of which were adopted from the literature. Some works have also adopted the power-gating technique to reduce static power. X-CGRA [Akb+20, OAk+18], PX-CGRA [Akb+18], GP-CGRA [Bra+18], and Quora [Ven+13] are the examples of these works which have gained energy-efficiency in the cost of area overhead and accuracy drop in the final output. The heterogeneous architecture of *GP-CGRA* consists of a 2D array of approximate and accurate tiles. The work has adopted a set of approximate adders and multipliers from the literature. The accurate divider is also adopted from OpenCores [Ope11] and further approximated based on the application requirements at *design* time. Based on the given quality constraints, various *feasible* CGRA configurations are generated through an iterative process by approximating an operation in the application's Data Flow Graph (DFG) and verifying whether the accuracy constraint is met. In the final configuration with the lowest power among the feasible ones, each CGRA tile is assigned a specific functional unit, which only supports one precision mode (selected at design time). Switching to accurate configuration is enabled at run-time via power-gating the approximate tiles. *X-CGRA* is composed of Quality Scalable PEs (QPSEs), each of which consists of accuracy-configurable approximate CLA adder and Dadda multiplier. Each QSPE supports four accuracy modes for the pair of adder and multiplier, i.e., each can be accurate or inexact. As only one approximate mode is supported for all the adders (and one for all the multipliers) of the whole CGRA, such *required* approximation level should be determined at design time. In X-CGRA work, assigning the quality level of each adder and multiplier node in the DFG representation of the application is solved via the ILP approach, which is rather a time-consuming heuristic. In another trend, the work of [Wan+22a] has exploited the SC to convert multiplication to simple bit-wise AND operation. However, such serial-based bit-stream processing has resulted in significant latency overhead. To amortize the area- and/or the latency-overhead of the aforementioned functional units, authors in [OR19] and our SIMDive work [EUK20] have narrowed their focus toward enabling *approximate* SIMD multiplier/divider. However, designing approximate SIMD/MIMD PEs for harnessing the DLP/ILP capabilities of error-resilient programs is still unexplored.

However, approximate CGRAs have two main shortcomings. First is the inevitable die-area overhead, which stems from collecting *separate* (and accuracy- configurable) adder, multiplier, or divider units. Moreover, supporting the heterogeneity of PEs, in an inter- or intra-PE manner, pose routing complexity and overhead to the overall CGRA [Bra+18, OAk+18, Akb+20]. These obstacles have hitherto prevented the offloading of division operations to most of the CGRA accelerators and relieved the host processor from context-switching on such occasional interrupts. Finally, studies like [HHS19] and our analysis (which will be detailed later on in this book) demonstrate the longer latency and higher energy of division operation compared to the multiplication, which can limit the operating frequency of the PEs. These obstacles have hitherto prevented offloading of division operations to most of CGRA accelerators and relieving host processor from context-switching on such occasional interrupts.

3.4.4 *Approximate Caches*

Approximating read and write operations STAxCache [Ran+17] employs two key approximation strategies for partial reads/writes and skipped writes. During partial reads, STAxCache ignores a few LSBs of each word in a cache block, setting them to 0 in the returned value. This approach bit lines and source lines corresponding to the LSBs to be gated for energy savings. For write operations, STAxCache introduces approximation through skipped or partial writes. It compares the incoming write data with the current cache contents and skips the write if the difference is within an acceptable threshold, retaining the stale value to save energy. If the write cannot be skipped, it employs a partial write strategy, ignoring a few LSBs as determined by the quality constraint.

Approximation of data values Doppelganger [Yaz+16] leverages value similarity among cache blocks to reduce the cache memory footprint. It identifies blocks with similar data and stores only one copy of the data, mapping multiple tags to that single entry. By storing only one representative block for multiple similar blocks, this approach trades off some precision for gains in memory efficiency and energy savings. The Doppelganger cache also utilizes an accurate cache to accommodate the accuracy-sensitive data, and the approximate cache is utilized to store only the error-resilient data. In the same context, the Bunker Cache [Mig+16] leverages spatial-value similarity—a phenomenon where data elements with similar values exhibit spatial regularity in memory, often being stored close to each other. The Bunker Cache maps similar data to the same cache location based solely on their memory address, reducing redundancy by modifying the cache indexing mechanism to group memory addresses with similar values. As a result, it decreases cache misses and energy consumption associated with cache accesses and off-chip memory transactions.

The Rollback-Free Value Prediction (RFVP) [Yaz+16] tackles two key memory bottlenecks: The bandwidth wall (limited off-chip bandwidth) and the memory wall (long access latency). RFVP introduces an approximate *value prediction* mechanism to address these challenges. When a cache miss occurs for safe-to-approximate load operations, RFVP predicts the requested values without verifying correctness or recovering from mispredictions. This avoids costly pipeline flushes and re-executions, allowing execution to continue without stalling for long-latency memory accesses, thereby mitigating the memory wall. Additionally, RFVP drops a fraction of cache-missing load requests after predicting their values, reducing memory bandwidth contention and addressing the bandwidth wall. The *drop rate* serves as a tunable parameter (approximation knob), enabling a trade-off between performance/energy efficiency and output quality.

3.5 Application and Software-Level Approximation Techniques

The techniques in this stream of research have targeted modifying the high-level implementation of applications. The broad spectrum of these techniques can be grouped into the following categories.

3.5.1 Data-Precision Reduction Through Precision Scaling and Truncation

Many literature studies have targeted (mixed-)precision tuning strategies for multi-kernel applications . However, they mostly focused on the NN domain (by e.g., exploring layer-wise quantization) [Rak+24, Rak+22, Jud+15, Mra+19, Sha+18, Las+19], and the investigations on other domains of applications has been limited. Overall, these works can be divided into two categories. (1) Either they adjust the precision of the individual operations (intra-kernel granularity) or (2) the precision of a group of operations as a kernel (inter-kernel granularity) in a multi-kernel application. As an example for the former, the study of [ED17] also considers SIMDization in its Word-Length Optimization (WLO) algorithm. The priority of blocks for SIMDization is based on their contribution to the overall execution time when evaluated in individual convolution, FIR, and Infinite Impulse Response (IIR) kernels. The work of [HYS20] has tackled the problem of WLO in cascaded image processing kernels (including filters such as de-noising, color conversion, and un-sharpening). The adopted search strategy of this study is a greedy heuristic that focuses mainly only on the quality constraint and afterward performs a local search to minimize other cost functions. In the follow-up work [HS21] the authors used Bayesian optimization which (1) optimizes functions that usually have no mathematical expression and/or derivatives and (2) is able to boost the search speed by avoiding local minima better.

As the pioneer of *bio-signal processing* approximation, XBioSip [PRS19b] has presented an approximation strategy for the Pan-Tompkins algorithm [JJ85] for heart-beat detection (which falls into the category of multi-kernel applications). In XBioSip, the precision of all kernels is uniformly tuned to 16-bit. Still, each kernel uses a different level of approximation (via truncation of LSBs) for its addition and multiplication operations. These levels are determined through an aggressive approach, i.e., kernels are approximated as much as possible in their appearance order. Therefore, this approach did not consider the significance of kernels in terms of their contribution to the *gained performance* versus *quality loss* when approximation is applied. Other works in this field have mostly focused on layer-wise quantization of NNs [LJG17, Tas+20, Jud+15, Mra+19]. In their adopted strategies, ranking the layers for quantization is mainly determined by their robustness to precision scaling. In other words, their proposed saliency metric is fluctuations in the final QoR ($\Delta\ QoR$). However, considering merely $\Delta\ QoR$, neglects the significance order of layers on the end-to-end gained performance, or in general, the *relation* between *performance gain* and *QoR loss*. Tackling this shortcoming, our proposed Plasticine methodology [Ebr+21] (discussed in Sect. 6.3) has proposed the $\frac{\Delta\ \text{Performance}}{\Delta\ QoR}$ as the saliency metric and then proposed a greedy heuristic for precision-tuning of multi-kernel applications, which has shown better results over state of the arts and provides near-optimal solutions. Considering the lower complexity and usually better run-time of greedy heuristics over multi-objective strategies such as Genetic Algorithms (GAs).

3.5.2 Computation Skipping and Task Perforation

These techniques can be effective for memory- and/or computation-intensive tasks with a broad variation, from general to more application-specific. From general techniques, one can denote loop perforation (i.e., loop iteration skipping), coefficient decimation, and increasing the stride step in the filters of various application kernels. On the other hand, pruning of neurons (individually or as a group, i.e., filter), early activation, and early classification are popular techniques for approximating NNs.

Loops are among the most resource-intensive and time consuming procedures in many applications. Therefore, skipping certain iterations (i.e., loop perforation) is investigated in many studies. For example, the authors of [Sid+11] have achieved performance gains of $2\times$ to $7\times$ with less than 10% reduction in output accuracy across various applications. Using a similar concept, the authors of [Chi+14] optimized the refinement algorithm of the K-Means clustering application by early terminating the iterations of the refinement algorithm in the clustering part if further iterations do not result in changes to the cluster assignments. Identifying such error-tolerant tasks/code segments of the application can be carried through the aid of compilers [Hof+09]. Pruning of neurons in NNs is another computation skipping technique, which can be applied structured (which removes entire units, e.g., neurons, channels, or layers along with all their connections) or non-structured (removing individual weights, based on their importance score, without regard to the network's structure) [Fan+19a, SSa+18, Han+18]. Defining Region of Interest (RoI) in images or reducing the image resolution are examples in the domain of image/video processing, which skips computations on the part of the pixels.

3.5.3 Data Reuse and Memoization

Memoization denotes storing the results of computation-expensive operations to be re-utilized in case of the same operations and input data recurrence. Approximate memoization extends this concept to produce imprecise results for new input data that exhibit some correlation with previously computed and stored data. For instance, [RBG13] leveraged the spatial locality of data in SIMD architectures. Similarly, [KKS15] has applied memoization to approximately similar input data for graphics applications. Coefficient repetition in image and signal processing kernels (and therefore using constant multipliers), weight sharing, and weight clustering [Xu+18, Dup+20b, Dup+20a] are other data reuse techniques.

3.5.4 Modifying Programming Language and Application Control Flow

The Eon framework introduced in [Sor+07] allows annotating the program's control flow with different energy levels. Hence, various execution paths with different energy levels can be selected dynamically at runtime to satisfy the overall user's energy constraint. The Topaz framework is another example [AR15], which first maps the error-tolerant computations to the imprecise functional elements and then identifies the outliers in the output using a quality control mechanism. Such outliers are then re-executed on the precise elements to satisfy the overall QoR. Finally, the EnerJ language [Sam+11] enables the annotation of data structures and their associated operations in two modes, i.e., accurate and approximate, where the approximate units execute the approximate variants.

3.6 Cross-Layer Approximation

Approximation techniques can be applied *individually* at different layers of abstraction (from application- to architecture- to circuit) or synergistically across layers (a.k.a cross-layer approximation [MSh+16]). The ultimate goal of cross-layer approximation is to apply a set of approximations at different layers of abstraction, with the goal of maximizing the performance gain while also not crossing a pre-defined accuracy threshold [Sha+16]. Herein, a compendium of the cross-layer strategies is presented, followed by pinpointing the relevant state of the arts in Table 3.2.

In the context of cross-layer approximation, most of the techniques adopted by the literature are from circuit- and application-level (for example, precision scaling and imprecise multipliers). In fact, most of the existing works have narrowed their investigation toward NNs [Mra+19, Fan+19a, SSa+18, Han+18]. Few other studies have also explored other domains of applications. For example, as discussed previously, XBioSip [PRS19b] has employed approximate multiplication and addition operations on precision-reduced kernels of heartbeat detection application. However, this work has uniformly set the precision of kernels to 16-bit. Therefore, not only is the effect of changing precision not investigated in this study, but also no sensitivity analysis is provided in that work. In fact, the kernels are approximated to their highest tolerance level, one after the other, in their normal appearance order.[3] In other words, they are approximated individually and as much as possible in the order they appear, without providing a saliency measure to rank the significance of kernels. As also admitted in [PRS19b], such a strategy achieves performance gain through approximating primary layer(s). This not only neglects the order of stages in terms

[3] In each kernel, a different number of LSBs are approximated using inexact instances of 1-bit addition and 2×2 multiplication operations.

Table 3.2 Summary of cross-layer approximation approaches[a] in multi-layer/kernel applications

Techniques	Description	Strategy	Constraint
Quant. (no cross-layer)	Layer-wise quantization via greedy heuristic [Jud+15], used in many state of the arts	Greedy	Layers significance only determined by accuracy loss
Inexact MUL, quantization	Select MULs wrt layer/weight significance, uniform 8-bit quant. nn [Tas+20]	Greedy	Layers significance determined only by accuracy loss
Inexact MUL, quantization	Layer-wise weight tuning w.r.t few MUL structure, in 8-bit quant. nn [Mra+19]	Genetic Algorithm	Significant exploration time
Inexact MUL, pruning	Neuron/layer pruning followed by weight tuning, uniform quant. nn [Sar+18]	Undefined	No detail for semi-greedy approach is discussed
MUL, pruning, quantization	Replacing MULs in half of neurons, uniform quant. nn [Han+18]	L2 norm	No justification for L2 norm/selection of Mul
MUL, prec. scaling	Mathematical modeling of inexact MULs & estimate their error on the final QoR of applications using ml methods [USK21]	Polynomial regression & multi-objective Bayesian optimiz.	Evaluated on one simple test, i.e., Gaussian smoothing filter
ADD/MUL, prec. scaling	Replacing LSBs' A in DD/MUL ECG kernels, fixed to 16-bit precision	—	Kernels approximated aggressively, in appearance order
MUL/DIV, prec. scaling	**Minimally-biased MUL/DIV, followed by (mixed) precision scaling of kernels**	**Greedy**	**Greedy heuristic → no guarantee for optimal solution**

[a] Valuable works like [LJG17] assess voltage scaling instead of inexact units (not in a cross-layer approach), hence, are not in the scope of this book

of sensitivity (intolerance to approximation), but also disregards their contributions in the total application resources.

The focus of other cross-layer designs has been mainly narrowed to the approximation of NNs [Mra+19, Fan+19a, SSa+18, Han+18]. In their adopted layer-wise approaches, ranking the layers for quantization has been determined merely based on their robustness to precision scaling [Jud+15, Mra+19]. Such a strategy neglects the significance order of layers on the end-to-end gained performance (when approximated by the same technique), or in general, the *relation* between *performance gain* and *QoR loss*. In their adopted layer-wise approaches, ranking the layers for quantization is determined merely based on their robustness to precision scaling [Jud+15, Mra+19]. Such a strategy neglects the significance order of layers on the end-to-end gained performance (when approximated by the same technique), or in general, the *relation* between *performance gain* and *QoR loss*.

The remaining studies in the literature have replaced an arithmetic unit with the inexact one, mainly in a single kernel of multi-kernel application, without exploring the effects of *changing* kernel's precision for a cross-layer approach. Recently, the state of the art CLAppED [Ull+22] has presented a cross-layer framework that uses a polynomial regression-based method to model approximate arithmetic operators and efficiently correlate approximate operators with their impact on an application's output quality. Therefore, CLAppED has incorporated the joint analysis and tuned the Degree of Freedom (DoF) across layers (inexact multipliers at the circuit-level and precision scaling at the application-level). It should be noted that managing the *adaptable* precision scaling is an application-level technique [LLS20, Han+18, DGK20b, DGK20a, BMS19] (such as Plasticine [Ebr+21] and state of the art IBM RAPID chip [Ven+21a, Ven+20], both of which support precision scaling technique in a multi-precision, SIMDized fashion), and different from non-adaptable truncation of operands (which cited as architecture- or circuit-level technique in studies such as [Yin+18].

Overall, cross-layer works suffer from a key shortcoming: Ignoring the fact that application-level optimizations will be poorly reflected or even throttled in non-optimized hardware (e.g., under-utilization of FPGA DSPs in precision scaling to low bit-width). Particularly, the latter has raised the quest for *Cross-Layer Approximation*, which aims to unlock the potential for approximation efficiently across layers of abstraction [Sha+16]. Although such valuable studies have attempted to enable this cohesive, cross-layer hierarchy by leveraging circuit-level imprecise adder and/or multiplier in tandem with application-level precision scaling, their evaluations have been mainly limited to NNs. Moreover, such approaches have left architecture-level opportunities untouched.

3.7 Error Sensitivity Analysis for Multi-kernel Applications

3.7.1 Static Analytical Error Estimation Methods

For evaluation of heuristic-generated solution (beside simulating them) different mathematical error estimation models have been employed. These models are based on Interval Arithmetic (IA), Affine Arithmetic (AA), Probability Mass Function

(PMF), variance propagation, or estimating a bound of mean and worst error metrics [Cas+20, LG19, HL11, Sen+19, Cha+13]. IA assumes complete independence of all variables, while AA captures some of the correlations between variables using a more complex form, reducing the pessimism. Specifically, IA derives guaranteed, yet pessimistic error bounds by propagating intervals which overestimates the error bounds. To overcome such pessimistic bound, AA has then proposed. The asset of AA over IA is that the former also considers the first order dependencies between variables, leading to less conservative error bounds. Nevertheless, the asymmetric nature of such error profile still induces poor results. Therefore, more recently modified Interval (MIA) and Affine (MAA) Arithmetic have been proposed that deal with the propagation of the error PMF.

Shortcomings of Error Estimation Methods: In general, three drawbacks are attributed to the mathematic-based approaches. First, the error expressions used for the propagation of mean or variance of error are derived using Monte Carlo simulations, assuming that inputs or error are uniformly distributed, or the errors are propagated through the system independently and do not interfere with each other. However, this assumption of uniformity is not guaranteed for different kernels and/or inexact variations of components. In fact, uniformly distributed lower-order bits may only be justified for the primary inputs, as in many approximate components, neither the output nor the error is uniformly distributed. The asymmetric structure of kernels in terms of connection of operations also raises more inaccurate prediction. Second, due to their complex nature, such mathematical estimations have been mainly limited to cascaded adders and/or multipliers in a single and rather small kernels such as Sobel or Gaussian filters, or DCT [VMB17]. Third, final application-level accuracy metrics, e.g., QRS for heartbeat do not directly correlate with noise power and there is no straightforward link between such error statistics and the formulation of the ultimate QoR. Finally, these models make heavy use of analytical models, making it difficult to scale for larger kernels having different approximate operators. For example, the number of terms required to propagate the PMF suffers from range explosion, even for small number of cascaded operations and reported to be prohibitively time consuming in case of large bit-widths inexact operators [Bon+18]. Therefore, the most reliable method for QoR evaluation is to perform a functional simulation of the application. In this context, a light-weight heuristic plays a pivotal role to address the run-time changes in the edge devices in a real-time.

3.7.2 Error Injection Methods

Performing a kernel-wise sensitivity analysis which reflects the effect of various approximation techniques on the end-to-end performance metrics of the multi-kernel application is a pre-requisite for an efficient cross-layer methodology. The adopted approaches for sensitivity analysis of error-resilient applications fall into two major categories: Application-specific or based on error-injection. In the following these studies are discussed chronologically.

Error-injection approaches ARC presented initially in [Chi+13b] and further employed in [Asc+20, Chi+13a] distinguishes the error resiliency of applications' innermost loops, considering that they consume ~70% of overall time in RMS applications. This resiliency identification is achieved through injecting random errors based on the Statistical Approximation Model (SAM) to the internal variables of loops and monitor the fluctuations at the application output. In SAM model, errors are selected from a normal distribution considering three parameters: Average error, error predictability, and error rate. Adaptive Statistical Approximation Model (ASAM) [GK17] has improved the ARC framework by introducing the number of approximate iterations as an adaptive parameter and divided the execution cycles of a radio astronomy workload into exact and approximate iterations. The authors have shown that this partitioning provides better results than ARC (in their radio astronomy case study). ASAC [Roy+14] discovers the amenability of variables to approximation through randomly applying bit-flips in 16 LSBs of the target variable and scoring the application output fluctuation. Their approach is reported to miss-classify non-approximable as approximable variables in some cases [Mit16]. This is while Approxilyzer [Ven+16] quantifies the quality impact of a single-bit flip in operand registers (both source and destination) of dynamic instructions. PAC [RWW15] (same authors as in the ASAC) upgrades ASAC methodology through assigning a Degree of Accuracy (DoA) required to satisfy a quality function. The DoA value in the PAC approach was translated to the number of bits which can be approximated. The authors assumed variables with DoA less than 0.5 (which causes the QoR loss of at most 3.4%) are approximable. Finally, the Intel's Approximate Computing Toolkit (iCAT) [Mis+14] analyzes error-resiliency by applying precision reduction on FP instructions assuming that the programmer (1) annotates the approximation amenable functions with high-level pragmas and (2) provides a quality function for the program. iCAT also supports different error patterns with different probability, operand bit-position, or bit-width.

Application-specific approaches The authors in [Rah+15] present two techniques for identifying approximable computations for a reduce-and-rank kernel e.g., K-Nearest Neighbor (KNN), which selects the top-K closest reference vectors to an input vector. Their *interleaved reduction and ranking* technique decomposes vector reductions into multiple sequential stages and interleaves them with the ranking steps. Then, based on intermediate results of these ranking steps, this technique identifies whether a particular vector is expected to appear in a final subset and future computations may be relaxed accordingly. The *Input Similarity based Approximations* technique leverages the temporal or spatial similarity between current and previous input to approximate or entirely skip parts of the computations. The applied approximation techniques are precision scaling and loop perforation. They also design a run-time framework for dynamically adapting the parameters of their two techniques for minimizing energy consumption while meeting the QoR target. Nogues et al. [NMP19] decompose the High Efficiency Video Coding (HEVC) decoder application into processing blocks and identifies sensitivity of each block to two classes of techniques, i.e., processing-oriented (skipping the filter tap and less accurate block) and data-oriented (fixed- rather than FP and bit-width optimization).

Targeting multi-kernel applications, most of the research has focused on the domain of NNs. Hanif et al. [HHS18b] have evaluated the effects of quantization and imprecise addition/multiplication in convolutional layers individually and observed the effects on the image classification accuracy. This is while works such as [SGA18, MSM18, WSX19] and their references identify less critical neurons based on e.g., Taylor decomposition [Mon+17] or partial derivative on the loss function at the output of each neuron. In short, the random error injection mechanisms cannot comprehensively characterize the error resiliency. Furthermore, the above-mentioned methods mostly categorize the variables either as approximable or sensitive without showing the end-to-end performance-quality trade-off of each kernel individually in the multi-kernel application. This can also highlight the priority of the kernel for approximation. This shortcoming is addressed through the proposed end-to-end, kernel-wise sensitivity analysis in Sect. 6.2.

3.8 Design Space Exploration Techniques

A huge number of application configurations will be generated by adjusting the approximation knobs in different kernels of a multi-kernel application. In fact, the number of configurations grow exponentially w.r.t the number of techniques, knobs, and application kernels. Since each of these configurations renders a unique trade-off between accuracy and performance gain, an efficient DSE heuristic is required to rapidly find the optimal or near-optimal configurations (i.e., having the highest performance gain and lowest accuracy drop) and prevent an exhaustive search. Therefore, herein, the focus is on the search strategies that attempted to find such Pareto or near-Pareto configurations. In brief, mainly four classes of heuristics have been utilized for such a DSE problem: (1) GA based approaches [MSV20, SVM19, Nep+19, HLR12, VM17, HMV20] expand the search space by generating a population of configurations in each iteration based on crossover and mutation actions. Besides having high computational complexity, the quality of later configurations in this class of heuristics depends heavily on the initial generations. However, the use of GA in this field has been limited to gate-level for building multipliers or the small image-processing filters. (2) Simulated annealing [Lee+16, Wit+19] randomly samples an action that improves the target metric while other actions are also accepted based on a predefined probability. Such algorithms are known to be time-consuming due to the large number of generated configurations. (3) Greedy strategies have less complexity and usually better run-time than multi-objective GAs but they have mostly considered ΔQoR (not the relation between ΔQoR and ΔPerformance-gain) as their saliency metric [LJG17, Tas+20, Jud+15, EK21] for their selection strategy. (4) Recently, some ML techniques have been adopted by researchers to reduce the exploration time of GA-based approaches. For example, the work of CLAppED

[Ull+22] has utilized Bayesian optimization, which is more suited for optimizing complex objective functions while reducing the evaluation time of each generated configuration [Moc74, Gal+20]. The search strategy in CLAppED framework is based on a multi-objective optimization method that uses the input configurations of the approximate multiplier as the design variables for a bi-objective optimization problem. Separate probabilistic models are then used for each design objective (performance gain and ultimate application accuracy). Every iteration of the Bayesian optimization is divided into three stages. First, an objective function is generated. Then, a set of inputs is chosen to train the model in the second stage. The training data is used to design the Surrogate function (a probabilistic prediction model). Finally, the acquisition function uses this prediction model to generate new candidate configurations to be evaluated by the objective function in the next iteration. This process is repeated for multiple iterations until the final design points are generated. However, the complexity of this ML-based method is high and, therefore, cannot be easily implemented in resource-constrained edge devices. Moreover, the high energy demand of such ML models might hinder their practical deployment for various dynamic exploration and management scenarios, which require run-time modification of the model.

In Sect. 6.3 our greedy-based heuristic is presented which attempts to address the aforementioned challenges, i.e., swiftly find Pareto- and near-Pareto configurations with lower complexity than e.g., GA- and ML-based models.

3.9 Distributed and In-Network Computing

3.9.1 In-Network Computing

Studies in this stream of research have targeted the deployment of *individual* kernels from various applications to the network [KT23, Zhe+23b]. The investigated domains can be summarized in ML [Zhe+23b], data aggregation [Sap+21, Lao+21], network management [Jin+17, LHC19], and image processing (edge detection [Gle+19]). In particular, the ML studies have shown promising results for tree-based classification models such as Decision Tree (DT) and Random Forest (RF) [Xav+22, Xav+21] or less complicated models such as Support Vector Machine (SVM), and K-Means. However, running resource-hungry models such as NNs on network switches is an arduous task and faced with practical deployment challenges, especially in ASIC-based switches [SSB18, Luo+22, Lui+21, Saq+21, Sir+22, XZ19, SB18]. In fact, even a heavily-quantized binary NN (BNN) with two layers of 64 and 32 neurons already exhausts the resources of an Intel Tofino switch [SB18].

3.9.2 Distributed INC

Among INC studies, recently, some works have attempted to partition networking applications implemented in P4[4] into kernels and distribute them over a chain of heterogeneous nodes. ClickINC [Xu+23] attempts to reduce the search space for the task placement problem by (i) grouping functions into blocks to be mapped together (e.g., within a loop) and (ii) grouping the heterogeneous platforms into fewer classes. Flightplan [Sul+21] disaggregates P4 kernels into segments *manually* and then combines graph-based and formal methods to solve the distribution problem. In fact, Flightplan performs an exhaustive search to find near-optimal solutions that minimize the allocated resources on heterogeneous hardware. To reduce the search time, DINC [Zhe+23a] adopts a Multi-objective ILP optimization strategy, where the objective function is a weighted linear combination of the execution latency (and disregarding the transmission latency) and the resource consumption required for *executing* the task. Finally, Hermes [Che+22b] targets minimizing the communication overheads in such inter-switch coordination scenarios by considering the per-packet byte overhead and the per-packet transmission latency. To solve this multi-objective optimization, the authors first formulate it to mixed-ILP and then solve it via a greedy-based heuristic. Despite these cutting-edge frameworks' valuable and initial efforts, they still neglect the error-resiliency potentials of applications in their analysis.

3.9.3 FPGA- Versus ASIC-Based Acceleration in INC

Computations in the network devices are performed either through applying a chain of match-actions—implemented via ALUs and Match-Action Tables (M/As or MATs)—in ASIC [Geb+20], or LUTs in FPGAs. Despite the higher packet-processing speed and power-efficiency of ASIC switches compared to FPGA counterparts, ASICs still suffer from multiple shortcomings. For example, the lack of support for *computing* operations (e.g., multiplication/division of arbitrary operands [MSV22] that are exhaustively used in ML and matrix-multiplication based applications). Moreover, the fixed number of pipeline stages, along with the limited on-switch memory capacity in a typical ASIC switch (a few 10s of MB of SRAM [Int22, Int18]), leads to a narrow range of custom functions that can be implemented via M/A tables [XZ19]. These deficiencies, along with the expensive hardware update in ASIC devices, have restricted the deployment of complex algorithms to them [Zhe+22a, Beh+20, Win+20, Zho+20]. Although some studies have also tried to modify the ASIC data plane or design dedicated modules [Swa+22] to alleviate such limitations, all of which are experimental and not off-the-shelf solutions that can be easily adopted.

[4] P4 is the de-facto language for programming network data plane [Bar20, Net23a, Wan+17, Iba+19].

In fact, major telecommunication enterprises [Bob+22, Wan+22b, Jin+20, Wit+20, Nok19, Joh+20, EL20, Nie+20, Cis22, Ji+13, Fra18, Net23b] already rely on broad utilization of FPGA in the edge-to-cloud continuum (e.g., switches, routers, base stations, NICs, and NPUs [Xil23, All+23, Xil20, Adv24b, Int19, Int23a, Int21, Intar, Nap18, AMD22, NVI24b, NVI24a, Fir+18, Gup15, Net18, Luo18]).

In fact, major telecommunication enterprises and cloud service providers already rely on broad utilization of FPGA in the edge-to-cloud continuum, including switches, routers, base stations, NICs, and NPUs [Bob+22, Wan+22b, Cis22, Jin+20, Wit+20, Joh+20, EL20, Nie+20]. In particular, FPGA is used as a co-processor to accelerate various tasks, from control/traffic management and load balancing to co-processing statistics and cryptography/security analysis [Iba+19, Bre+20]. The vast deployment of FPGA is actually endowed with their multiple advantages over ASICs: (1) FPGAs are not only well suited for packet processing with different protocols [Fie+16] but also enjoy a post-fabrication hardware data-path versatility along with a faster prototyping and lower NRE cost. (2) Through enabling high parallelism, FPGAs are suitable hardware platforms for the implementation of, e.g., matrix-multiplication-based applications. Such feature has enabled up to an order-of-magnitude (or more) improvement in throughput and/or latency for a variety of in-network accelerated tasks [CF20, Qu+15, Bre+20, Tok+19, TMZ18, FVS15, Gob+21]. (3) Although integrating an FPGA as the co-processor for the switches may increase the power consumption of a *single* switch, the overall energy or *performance per Watt* will be enhanced for certain application kernels [Bre+20, Tok+19, TMZ18, FVS15, Gob+21], when compared to network-deployed Central Processing Units (CPUs), Graphics Processing Units (GPUs), and General Purpose Processors (GPPs). These reasons have encouraged the in-network acceleration of many tasks with FPGA [Zil+14, Iba+19].

3.9.4 Approximation in INC

To cope with the implementation limitations in ASIC switches and in general, P4 language [Bos+14], some works have adopted approximation techniques to reduce the complexity of computations. These works are classified and discussed below.

Approximation of Data Type The lack of support for FP operations in ASIC switches and P4 [Ope22] has pushed the designers to approximate their computations with the integer or fixed-point precision [Pat+22, NWB22, XZ19, LAG19]. In fact, this issue is also a major barrier to performing the training of ML models on the network nodes.

Approximation of Primitives/Individual Kernels Researchers have attempted to approximate widely-used primitives (such as multiplication [Sha+17, Zhe+22b, Zhe+22a, Zhe+23b], logarithm [DSS20], and trigonometric functions [Kun+21]) using either dedicated LUTs/MATs or convert them into shift/addition operations. However, such approximations are applied arbitrarily to the applications whenever possible, while a systematic error-sensitivity analysis was never made. Therefore, the applications' ultimate QoR were sacrificed up to 10%, only by adopting a single technique (i.e., inexact multiplication in [Sha+17]). Moreover, pre-computing the approximated values and storing them in dedicated memories lacks scalability to be applied to more complex functionalities.

Approximation of Multi-kernel Applications state of the arts in this line of research have only targeted NNs and applied extreme quantization,[5] i.e., substituting the multiplication operation with XNOR. Although modern SmartNICs, which benefit from more resources compared to ASIC programmable switches such as N3IC [Sir+22], have been able to implement, e.g., a 3-layer BNN, they are physically deployed only at specific locations within the network [AGF+23].

3.9.5 INC-Accelerated BSS

Aside from applications in ML domain, most of INC works have targeted the in-network acceleration of BSS program [Ces+20, Rüt+18, Wu+23],[6] due to two reasons. (1) BSS is a common task in various signal processing domains from audio to image and bio-signals [CJ10]. Specifically, acoustic BSS is crucial for speech recognition, natural language processing, voice assistants, etc. Moreover, with the emergence of smart factories within the era of 5G, acoustic BSS also plays a pivotal role in quality control and the automated supervision of production lines (to detect the malfunction in faulty products at early stages, prevent the product failure, and thereby, decrease the downtime and maintenance costs, dubbed as predictive maintenance). (2) Performing controlling tasks such as BSS in real-time is paramount in latency-critical applications (especially anomaly detection and decision-making) so that the necessary responses can be triggered prior to deterioration. However, it has been shown that the performance of a centralized cloud-based approach for BSS is bounded by the data collection and transmission time. The long data transfer time may delay the prompt actions in the mission-critical services in the upcoming 6G era [Rüt+18]. Hence, performing this task in real-time carries great potential. Tackling these issues, recent studies have attempted to accelerate this application by offloading it to the network to reduce its Mean Time To Respond (MTTR). The implementation of BSS in these studies is either via a NN-based [Wu+22c, Wu+20b, Wu+21d] or an algorithmic-based approach [Wu+23, Wu+21c, Wu+21b]). Despite the effort of these pioneering studies, there exist key drawbacks that hinder their practical deployment in an INC setup: (1) The adopted models, such as ResNet [He+16] and MobileNet [How+17] are resource-intensive and hence, not easily acceleratable on the resource-constrained network nodes. Moreover, there exists many complex operations in the structure of BSS application that are not directly supported in P4

[5] Quantization is a precision scaling technique applied in the NN domain.

[6] The authors of [Wu+23] have also published other INC-tailored implementations for BSS application in [Wu+22c, Wu+20b, Wu+21d, Wu+21c, Wu+21b].

(see Sect. 7.3.1). These issues have been neglected in all previous studies. (2) It is hard to obtain enough labeled audio data for model training [Wu+23, Wu+22b]. (3) Updating and re-training the NN models is costly, especially after being deployed on network nodes.

The main idea behind the second category (implemented based on the Independent Component Analysis (ICA) algorithm, i.e., FastICA [Hyv99]) is to provide an early separation result and then *progressively* optimize its accuracy by iteratively increasing samples over the chain of hops along the network forwarding path. Although these works realized the drawbacks of the NN-based approaches and looked for alternative decentralized/INC-oriented strategy, they still suffer from three shortcomings: (1) If the threshold for the gradient of improvement—over consecutive hops—is too loose, the construction of separation matrix will terminate prematurely, and if too tight, it may never terminate [Wu+20a]. (2) The complexity of the algorithmic solutions is still very high. For example, these approaches adopted the FP precision for the computations, which is not only unsupported in P4 but also is costly in terms of resource footprint (see Sect. 7.3.1). (3) The computation time for such high-precision solutions is still several magnitudes larger than the usual packet-size processing time [Wu+23]. Moreover, the computation itself is further delayed due to caching the new data at *each of* the intermediate network nodes. These issues have increased the application response time and confined the performance gained from acceleration through the network. In fact, as also admitted by the authors in [Wu+20a, Wu+21c], the processing time would be worse in some extreme cases (a large number of hops) than the centralized approach. To the best of our knowledge, this is the first work to investigate the utilization of approximation potentials for the in-network acceleration of multi-kernel applications, with the aim of enhancing performance, such as improving energy efficiency or reducing application response time.

In X-DINC (Chap. 7), the algorithmic approach is targeted and the aforementioned issues are addressed by (1) tailoring the computational structure of such multi-kernel applications for an INC setup and (2) aiming an end-to-end performance gain in a distributed implementation approach, by capitalizing on the error-resiliency of these applications and other optimization opportunities, in a network hierarchy.

3.10 Conclusion

This chapter discussed the related work and the widely-used approximation techniques in both data acquisition and processing/computing phases of an application. Approximate sampling and data compression are the techniques used for the data acquisition phase, while a broader range of techniques are proposed for the processing/computing phase (which span circuit—to architecture—to application layers of the computing stack). Afterward, a summary of cross-layer strategies was provided, along with error-sensitivity analysis techniques and DSE heuristics, which are prerequisites for an efficient cross-layer methodology. Finally, the state of the art studies in the domain of INC were presented: The distribution approaches, basic approximation approaches, FPGA versus ASIC-based acceleration. Although these works targeted enabling the in-network acceleration of applications on resource-constrained network elements, they fall short on presenting a cross-layer strategy to efficiently apply a a chain of approximations across the computing stack in a network hierarchy. In fact, none of the cutting-edge studies has identified the potential knobs or proposed a systematic approach to reveal the error-resiliency of 5G/6G applications to different techniques. We aim to address these issues in this book.

Chapter 4
Circuit-Level Approximations

Abstract This chapter motivates and outlines FPGA-tailored approximate arithmetic units for efficient multiplication and division in signal-processing and machine learning workloads. The chapter targets multiplication and division because they dominate energy and resource usage, and existing LUT-based soft cores on FPGAs suffer from high latency and large area despite being preferred over DSP blocks for low bit-width operations. It introduces new approximate multipliers and dividers built on Mitchell's logarithmic algorithms, optimized for high throughput and low area/energy via fine-grain pipelining and SIMD/MIMD execution styles. The designs are customized for FPGA fabrics and deliver accuracy comparable to, or better than, state-of-the-art ASIC-oriented units, while incorporating distinct approximation and implementation innovations. The chapter thus positions these units as practical building blocks that reduce the cost of deploying off-the-shelf FPGA for demanding signal-processing and machine learning applications. It concludes by detailing the specific technical contributions of each proposed design.

Keywords Approximate computing · FPGA · SIMD · MIMD · High-throughput · Pipeline · Energy-efficiency · Area optimization · Mitchell's algorithm · Logarithmic hybrid multiplier-divider

4.1 Introduction

This book primarily focuses on designing approximate configurations for multiplication and division, since they are the most widely used and energy-hungry arithmetic operations in numerous applications, including signal-processing and ML domains. As discussed in Sect. 1.5, to enable the high-speed execution of multiplication and/or division on FPGAs , designers are compelled to employ soft IPs for these operations [Xil15, Xil16b]. In fact, exploitation of soft IPs instead of DSPs for low bit-width operations has been suggested by academia and industry [Xil18a, VL20]. Nonetheless, the long latency and high resource footprint of LUT-based IPs still needs to be decreased to facilitate the adoption of off-the-shelf FPGAs for implementing signal-processing and ML-based applications.

Z. Ebrahimi and A. Kumar, *Cross-Layer Approximation and In-Network Acceleration*,
https://doi.org/10.1007/978-3-032-21712-7_4

This chapter presents various novel approximate multipliers and dividers targeting high throughput and low area/energy. Our approximate designs utilize Mitchell's algorithms as the base for designing various multipliers and dividers. To reach a very high throughput, the structure of proposed functional units is further customized toward a fine-grain pipelining or SIMD/MIMD execution approach.

Although the error-reduction approaches of state of the art and proposed logarithmic multipliers and dividers in this book are similar in spirit, each has been designed for either ASIC or FPGA and targeted for a different design goal. First, while prior designs of MBM [HHS18a], INZeD [HHS18a], and REALM [HSa+20] are ASIC-customized multipliers and dividers, our LeAp, SIMDive, and RAPID designs implement the FPGA-customized versions with similar or higher accuracy. In fact, each of the proposed designs presents specialized novelties in both approximation and implementation perspectives. The novel contributions for each of the proposed multiplier and divider designs are listed below.

LeAp Non-pipelined Multiplier:

- LeAp presents a novel architecture for the approximate log calculator, which is customized specifically for FPGAs. In detail, while prior works used large priority encoders and barrel shifters for implementing the leading-one detection and fractional part alignment (log calculation), LeAp implements both steps concurrently with the same, small number of LUTs [Pra+18].
- The error-reduction scheme of LeAp not only has reduced the error to (ARE $< 1\%$ which surpassed the accuracy of MBM [HHS18a]) but also can be added to the original Mitchel's multiplier with a small overhead. In fact, unlike the error-refinement approaches at the time [HHS18a, LJ15]—which required additional circuitry for *adding* the error-refinement step—LeAp performed the *addition* of error-coefficients to the addition of fractional parts in the original Mitchell by directly configuring same LUTs and their associated fast carry chains.

SIMDive Hybrid Multiplier-Divider (SISD and SIMD/MIMD):

- SIMDive is the first hybrid approximate multiplier-divider in the literature, which can switch to either functionality on-the-fly, without the need for reconfiguration. Moreover, the SIMDive approach has significantly reduced the latency of division, up to $\sim 4\times$ compared to the accurate version.
- The novel error-reduction scheme of SIMDive is different from LeAp and is tunable to different accuracy-levels. In fact, SIMDive uses solely one 6-LUT for determining each bit of 64 error-reduction coefficients, the addition of which (to the original Mitchel design) is also performed with minimal overhead.
- To support data parallelism opportunities, SIMDive multiplier-divider is also designed with an SIMD approach (with the word lengths of 8-, 16-, and 32-bits) to speed the execution of vast domains of parallelizable tasks that can benefit from this feature. This is achieved by designing the sub-module with a hierarchical approach (e.g., 4-bit LODs) and connecting them with a minimal overhead. Interestingly, the SIMD units require a similar or smaller number of LUTs compared

to the accurate SISD version. To the best of our knowledge, no SIMD divider has been presented in the literature so far.

RAPID Non-pipelined and Pipelined Multiplier and Divider:

- RAPID has enabled multiplication with higher accuracy than LeAp and state-of-the-art Mitchell-based approaches (achieving the ARE $< 0.6\%$ through a limited number of coefficients). The errors are also minimally biased and centered around zero. A similar approach has also been adopted to design three versions of the divider. Although RAPID and SIMDive multipliers and dividers achieve a nearly similar accuracy level ($0.6 < \text{ARE} < 1$), their implementation is customized to be suited for different execution approaches, i.e., pipelining (RAPID) and SIMD (SIMDive).
- RAPID presents the first pipelined architecture for an approximate (integer) multiplier and divider design. The pipelined units are designed in 2- 3-, and 4-stage variations and achieve up to $8.1\times$ higher throughput and $6.8\times$ higher throughput/Watt over their accurate IP-based counterparts from Xilinx Vivado. These variations can be adopted w.r.t the target operating frequency of FPGA and application requirement.
- Contrary to the previous works, which evaluated their designs only in a single kernel of multi-kernel applications, the efficacy of the RAPID multipliers and dividers is appraised on the end-to-end implementation of three multi-kernel applications.

In the following, LeAp, RAPID, and SIMDive are introduced, with LeAp and RAPID grouped together since they employ a more similar error-reduction approach than SIMDive.

4.2 LeAp: Leading-One Detection-Based Approximate Softcore Multiplier with Tunable Accuracy

LeAp is the first approximate multiplier designed for this book, which targets area-, throughput-, and energy efficiency. The novel approximate log calculator in LeAp utilizes 6-LUTs and fast carry chains in a way to efficiently implement the Mitchell-based multiplier for an FPGA platform. The motivation behind designing LeAp based on Mitchell's algorithm is the translation of multiplication into addition in the logarithmic representation. This translation, in particular, enables resource savings for an FPGA, which is already equipped with fast carry chains that are hardened to accelerate the addition operation. Therefore, occupying fewer resources can provide the opportunity to execute multiplication-intensive workloads on a single chip. LeAp is proposed with three error-reduction schemes which reduce the ARE (PRE) of the original Mitchell design from 3.85% (11.11%) to 0.98% (4.7%). These error-refinement schemes are independent of multiplier size and pose marginal area overhead to the original Mitchell's design. In the following, the error-reduction schemes of the LeAp is discussed and then demonstrate and detail the hardware implementation of the LeAp design.

4.2.1 LeAp 2-, 3-, and 5-Coefficient Error-Reduction Schemes

Mitchell's error for 8×8 multiplier is plotted in Fig. 4.1 which is based on Eq. 4.1. The following three points can be observed based on this figure and equation:

$$E_P = P - \widetilde{P} = \begin{cases} 2^{k_1+k_2}(x_1 x_2), & x_1 + x_2 < 1 \\ 2^{k_1+k_2}(1 - x_1 - x_2 + x_1 x_2), & x_1 + x_2 \geq 1 \end{cases} \tag{4.1}$$

- Figure 4.1a shows that the error magnitude differs in each power-of-two interval. Hence, a single correction term added to the output will not fit all multiplier sizes.
- Equation 4.1 proves proportionally replication of error in each power-of-two. Hence, irrespective of k_1/k_2, a unique refinement scheme may fit all multiplier sizes, and it can be added to the fractional part before scaling to save more resources.
- Figure 4.1b demonstrates that error is non-uniformly distributed in the whole interval. Still, it is nearly symmetrical: For most input combinations, it tends to be near zero (at the beginning and end of each interval) and reaches its peak in the middle. Part (c) of this figure demonstrates that approximate products with the highest relative error are significantly less than others (especially cases with less than 1% error). Still, their magnitude is multiple times the average error throughout the whole interval, i.e., 0.0833, which is obtained from Eq. 4.2. Therefore, assigning merely a single reduction term (like [HHS18a]) to the whole interval is inefficient and results in output overflow.

$$\begin{aligned} Avg_{E_P} &= \frac{1}{(1-0)(1-0)} \int_0^1 \int_0^1 E_p dx_2 dx_1 \\ &= 2^{k_1+k_2+1} \int_0^1 \int_{1-x_1}^1 (x_1 x_2 - x_1 - x_2) dx_2 dx_1. \\ +2^{k_1+k_2} &\int_0^1 \int_0^{1-x-1} (x_1 x_2) dx_2 dx_1 = 0.0833 \times 2^{k_1+k_2} \end{aligned} \tag{4.2}$$

The insights obtained from these points and the overflow problem in MBM [HHS18a] (due to using only one error-reduction coefficient) motivate using multiple terms appropriately opted based on the fractional parts. Therefore, the squarish region between each power-of-two was divided into multiple regions, each with a distinctive coefficient representing the average error of the Mitchell algorithm for that region. Two factors have been tried to be optimized in each of the LeAp-2 to LeAp-5 schemes. First, the *error distribution* × *error* magnitude in each region has been aimed to be minimized (can be estimated as the integral of the error magnitude of the region). Second, the overhead of implementing the partitioning approach in hardware is aimed to be minimized. This is achieved by only analyzing four MSBs of fractional parts to select the appropriate error-reduction terms for each pair of input combinations.

LeAp 2-Coefficient Error-Reduction Scheme This scheme targets improved accuracy while enduring minimal overhead imposed by error-coefficient selection circuitry. To this end, the main squarish region was divided into two squares (Fig. 4.2a). For numbers bounded in the innermost zone with a higher relative error, the error-coefficient with a value of 0.0938 (obtained from a measurement similar to Eq. 4.2 for this interval) was applied. Analogously, the outer-zone coefficient is obtained in the

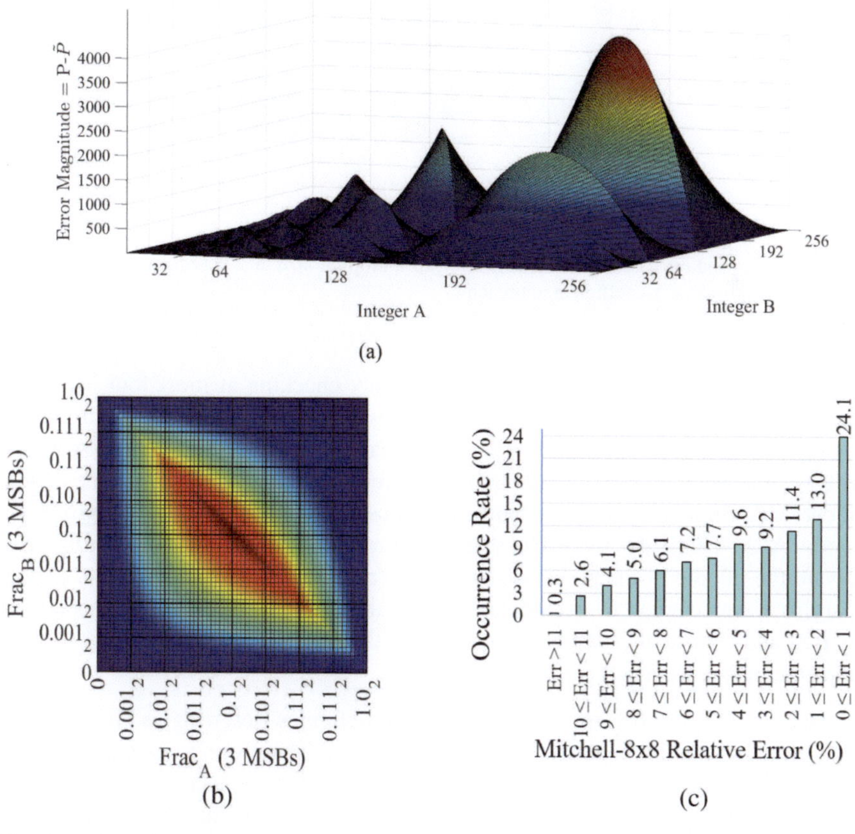

Fig. 4.1 Mitchell's error: **a** side view of error magnitude (8 × 8 multiplier), **b** top view of error (same for any size of multiplier), **c** relative error distribution

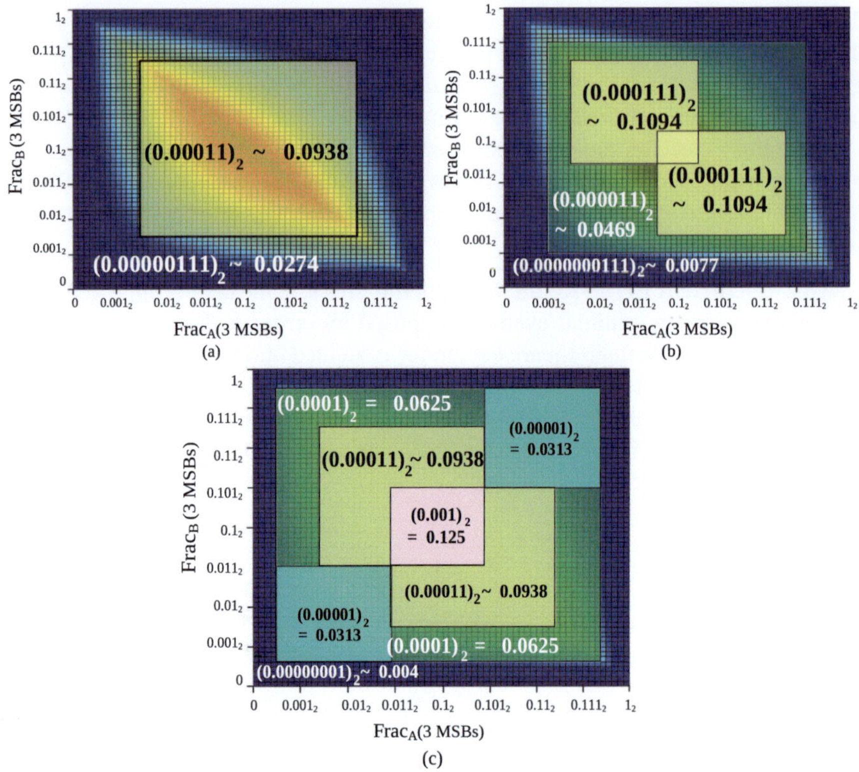

Fig. 4.2 Proposed error-reduction schemes for all multiplier size, **a** 2-coefficient approach **b** 3-coefficient approach **c** 5-coefficient approach

same manner. This partitioning is based on only checking at most four MSBs of fractional parts. Note that although the proposed partitioning is not optimum (as the highest errors are bounded in a rhombus shape), yet it is significantly rewarding in terms of accuracy with an endurable overhead. In contrast, the implementation of checking rhombus borders in hardware is costly and will not further improve the overall ARE. This is due to the fact that the number of high-magnitude errors reduces toward the middle of the power-of-two range (refer to Fig. 4.1c). Therefore, region margins were selected to eventuate in the best resource–error trade-off, obtained by trial and error from both behavioral in-depth error characterization in C++ and implementation reports from Vivado.

LeAp 3-, and 5-Coefficient Error-Reduction Schemes Referring to Fig. 4.1b, the high-error zone is not squarish but still symmetrical. Therefore, to gain higher precision, variants of partitioning with their associated resource usage were analyzed; among those, the hardware implementation of partitioning illustrated in Fig. 4.2b and c resulted in the smallest overhead and pruned other candidates (for the sake of brevity, 4-coefficient is skipped). In both approaches still, four MSBs of the fractional parts are analyzed but with a bit deeper conditional checking of borders. This is while in another work-study [LJ15], 4 up to 14 bits are checked, which its penalty would thwart the gain. It is noteworthy to mention another accuracy drawback (besides output overflow) attributed to the mechanism proposed in MBM [HHS18a]: There are three cases where the peak error is still equal to its counterpart in the original Mitchell, 11.11% and seven others higher than 8%. These cases occur when the summation of integer parts is small, and therefore, some bits of error-reduction term will not be shifted to the final output. The authors neither measured these errors nor presented a handling circuit; they simply suggested putting a comparator and avoiding adding error-reduction terms in these cases. In our schemes, such errors are alleviated by assigning appropriate coefficients in each scenario.

4.2.2 Hardware Implementation of LeAp

Figure 4.3a shows the LeAp multiplier design. In our design, the detection of the leading-one and re-aligning of the fractional parts are performed concurrently, via the proposed *approximate log calculator*, described below.

LeAp Approximate Log Calculator Each input in our approach is divided into 4-bit segments,[1] and the logical *OR* function is applied to the bits in each group to find segments having a bit with the value '1'. For instance, consider 16-bit input *A [16]* = *'0000000110101010'*. The leading-one position is located in *A [8]*. To find this position in our approach, first, the leading one was detected, which is located in the third segment from the right-hand side (A[11 down to 8]). Then, the decimal value 11 was loaded into the registers connected to their dedicated carry chain. Afterward, the position of the leading-one in the most significant group in was determined, at most, four cycles, using one FPGA slice, as shown in Fig. 4.3b: in each cycle, the value of the register was counted down and shift bits to the left by one bit until the leading-one is found (counting down from 11 to 8 takes four cycles).

From the implementation perspective, each slice in a Virtex-7 CLB can be configured as a downward counter,[2] as shown in Fig. 4.3b: Counting down the 4-bit register (needed in 16-bit LOD) is equal to adding it with binary "1111" and simply neglecting the overflow. In doing so, the carry chain output (a 4-bit register with a default value of 15) was connected to the first inputs (I_0) of four LUTs in the slice and assign '1' to I_1 (Fig. 4.3b). LUTs are configured to implement truth-tables of carry-generate (O_5) and carry-propagate (O_6) functions, so the associated carry chain implements a 4-bit downward counter. Therefore, as soon as the position of the leading-one is

---

[1] The segmentation granularity enables a trade-off in latency and energy, which could be a deterministic parameter for the designers in relatively large multipliers. It should be noted that the segmentation of 4 to 8-bit was analyzed for inputs, but for the sake of brevity, results for 4-bit grouping are presented, which renders the best energy-delay product.

[2] which consists of four 6-LUTs with their associated carry logic, a 1-bit shifter, and a register for saving the intermediate data.

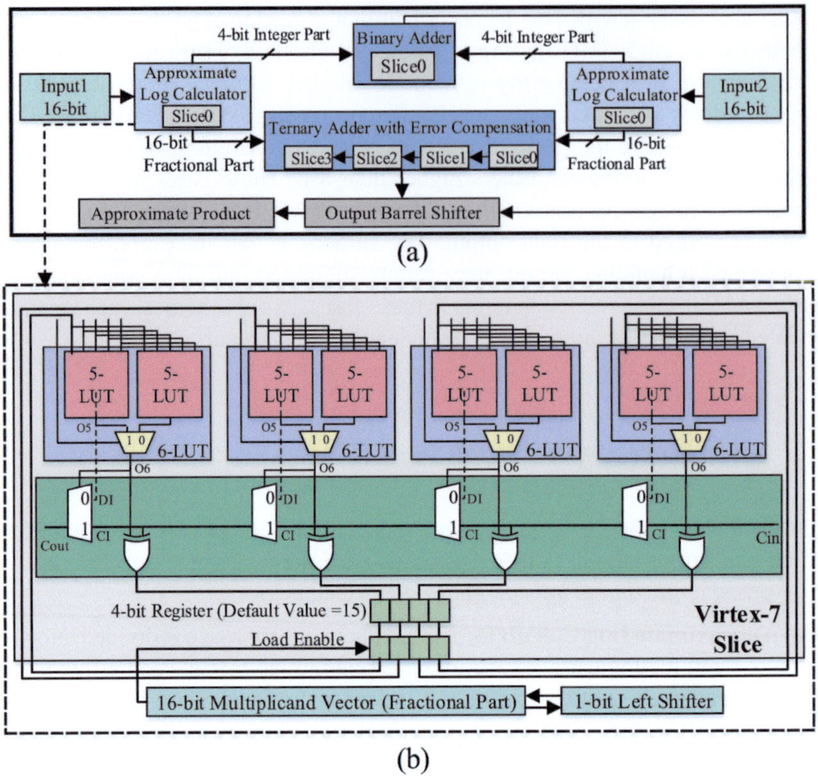

Fig. 4.3 16 × 16 LeAp architecture, **a** overall structure, **b** Virtex-7 slice (configured for approximate log calculator)

found, the fractional part (which has already been shifted to the left in each clock) is also re-aligned and located in the intermediate register.

Adding Error Coefficient to Approximate Log After determining the approximate log of inputs, adding integer parts together (each 4-bit) is fulfilled by one FPGA slice. Similarly, adding the fractional parts with the error-coefficient term (each 15-bit) is implemented with four FPGA slice. In fact, utilizing LUT primitives (from the Xilinx UNISIM library [Xil13]) is beneficial for us, as they can be configured to implement a ternary adder, which perfectly suits our proposed error-reduction method. Our approach combines the process of adding error-reduction terms with fractional parts within the same resources in a single step. This is particularly well-suited to our error-reduction approach, as it enables the combination of adding error-reduction coefficients with fractional parts within the same resource footprint, all in a single step. In fact, regardless of the ternary adder size and compared to the binary version, only one more bit at MSB position is needed, as $\text{frac1}_i + \text{frac2}_i + \text{error\_coefficient}_i + C_{\text{in}}$ (C_{out} from prior bit) may result in 3 bits, requiring another LUT at the end of the chain [Bar+20]. Therefore, no extra circuitry is needed to add the error-reduction term in our proposed schemes. Moreover, the delay of FPGA primitives is fixed. Therefore, adding the error-reduction term simultaneously as fractional parts does not impose additional overhead on the design (unlike [HHS18a] where addition is dependent on the carry out from summation of the fractional part, which lengthens design bottleneck path). The supplementary circuit in our design consists of registers for storing the error-coefficient terms (constant) and a selection module (multiplexer) that selects the suitable error-reduction terms for the pair of input operands.

4.3 RAPID: AppRoximate LogArithmic Pipelined MultIplier and Divider with Tunable Accuracy

The rapid updates in error-resilient applications and their quest for high throughput have motivated us to design fast and high-throughput approximate functional units for FPGAs. This Section presents the RAPID multipliers and dividers as the first pipelined approximate architectures customized for FPGAs. To implement Mitchell's approximate algorithms, the proposed units efficiently utilize 6-LUTs and fast carry chains. RAPID is the extended version of the LeAp multiplier with an improved error-refinement scheme that not only has higher accuracy over the baseline Mitchell's design but also is designed for a divider (no divider was proposed in LeAp) and boosts the accuracy of multiplication and division to 99.4% for arbitrary size of operands. This means that the ARE (PRE) of RAPIDâŁŒ are 0.6% (3.6%) for multiplication and 0.6% (3.4%) for division. In the following section, the error-reduction schemes of RAPID are presented, and the structure of the RAPID multiplier and divider architectures (non-pipelined and pipelined) is elaborated.

4.3.1 RAPID Error-Reduction Schemes for Multiplication and Division

Mitchell's error for 8-bit multiplication and division are (re-)formulated in Eqs. 4.3 and 4.4. Inspecting the behavior of the error has provided the following insights, some of which are also noted and endorsed by state of the arts [HSa+20, EUK20, HHS18a, HHS19]. Based on these observations, three error-reduction schemes have been proposed for each of the multiplication and division operations in Figs. 4.4 and 4.5, achieving higher accuracy than the LeAp designs.

$$E_P = P - \widetilde{P} = \begin{cases} 2^{k_1+k_2}(x_1 x_2), & x_1 + x_2 < 1 \\ 2^{k_1+k_2}(1 - x_1 - x_2 + x_1 x_2), & x_1 + x_2 \geq 1 \end{cases} \tag{4.3}$$

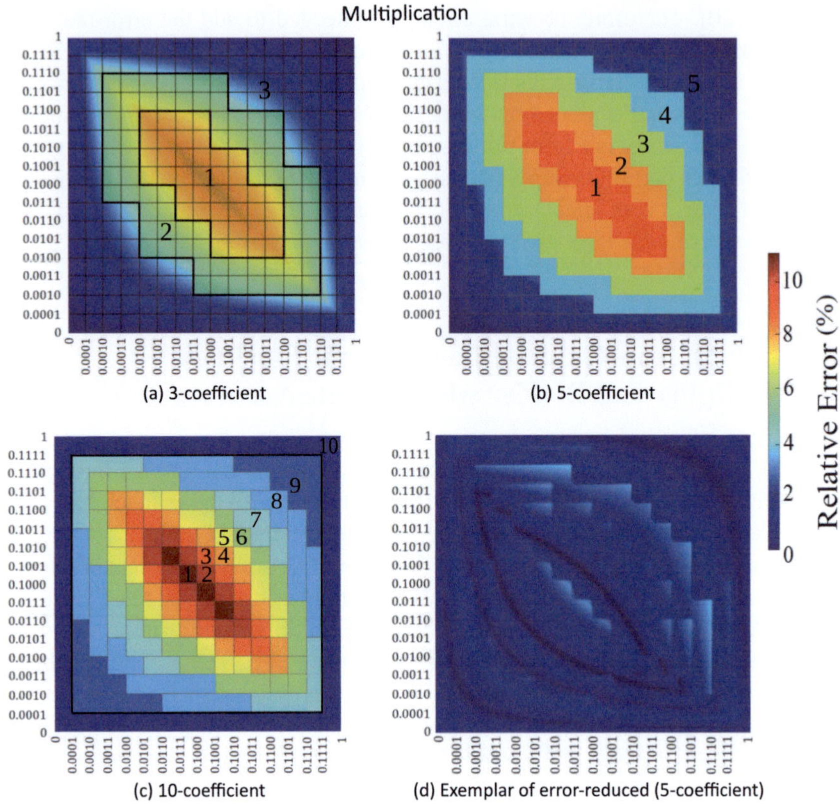

Fig. 4.4 Proposed error-reduction schemes of RAPID for multiplication, based on four MSBs of fractional parts

$$E_D = D - \tilde{D} = \begin{cases} 2^{k_1-k_2}\,\frac{(x_1(x_2-1)+x_2-(x_2)^2)}{2(1+x_2)}, & x_1 - x_2 < 0 \\ 2^{k_1-k_2}\,\frac{(x_1x_2-(x_2)^2)}{1+x_2}, & x_1 - x_2 \geq 0 \end{cases} \tag{4.4}$$

- As discussed earlier, these equations demonstrate the different error magnitudes in each power-of-two interval (also for division error). Therefore, merely adding a single correction term to the output (as proposed in INZeD [HHS19], and MBM [HHS18a]) is not efficient and results in many output overflow cases [HSa+20, EUK20].
- The equations also reflect the proportional replication of error in each power-of-two. This behavior is repeated for each size of multiplication or division (irrespective of k_1 and k_2). This repetition allows the same error-reduction approach to be applied to all multiplier or divider sizes. In fact, a set of coefficients can be added to fractional parts; before that, their summation becomes scaled [HHS18a, HHS19]).

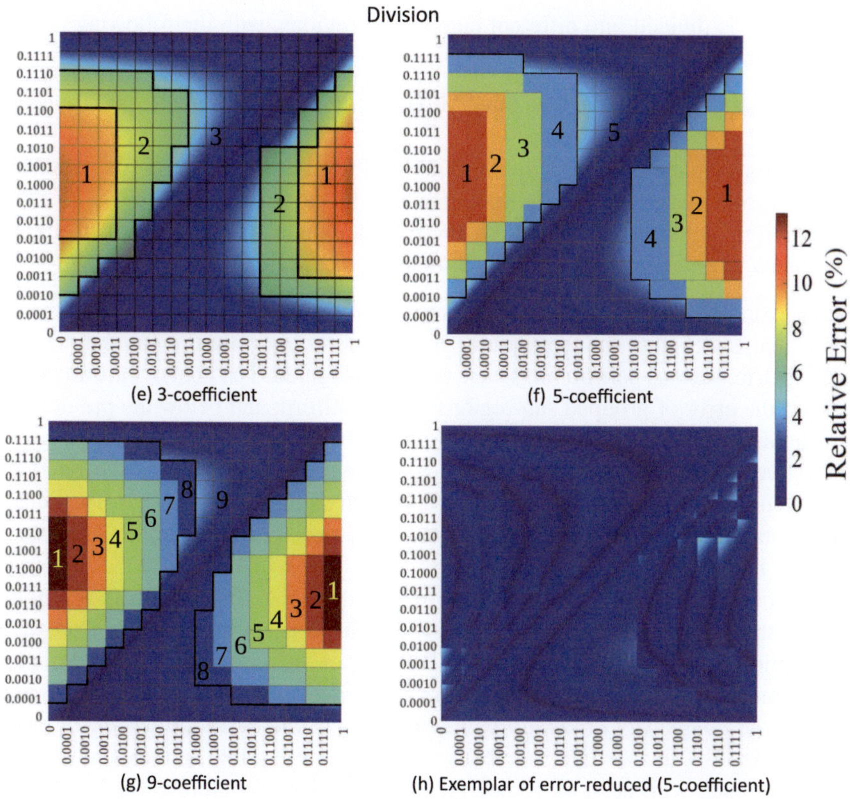

Fig. 4.5 Proposed error reduction schemes of RAPID for division, based on four MSBs of fractional parts

- As can be seen in Figs. 4.4 and 4.5a, and e, some of the nearby sub-regions have very small errors (less than 2%), while the changes of error occur is more pronounced in other sub-regions. This motivates grouping the regions with similar errors and assigning more coefficients to the sub-regions with higher errors (to reduce the error to a certain bound efficiently).

It is also worth noting that efficient partitioning should result in a small number of coefficients to minimize the overhead of error reduction. For example, considering even four MSBs of fractional parts—for partitioning—results in $2^4 \times 2^4 = 256$ coefficients (similar to approaches such as REALM [HSa+20]) and high resource footprint for implementing in hardware. This limits the scalability of such schemes. To cope with the overflow problem in INZeD and MBM and the over-provisioned number of coefficients (i.e., 256) in REALM, the insights obtained from the aforementioned observations incentivize reducing the error-reduction terms by efficiently clustering them in fewer groups. Afterward, the squarish region between each power-of-two pair is divided into different numbers of regions, with the following factors considered for partitioning:

(1) Considering four rather than three MSBs of fractional parts to increase the accuracy.
(2) Minimizing the number of sub-regions while considering four MSBs, to reduce the associated overheads for implementing the partitioning.
(3) Minimizing the error distribution × error magnitude in each region (can be estimated to the integral of error-magnitude for that region).

We have proposed three error-reduction schemes in Figs. 4.4 and 4.5, based on the in-depth analysis of error and the resource-usage reports from Vivado. To minimize the ARE metric, various partitionings have been investigated. The goal of partitioning is to keep the error of grouped sub-regions nearly uniform and below the pre-defined thresholds, i.e., $\sim 4\%, 3\%$, and 2.5% for 3-, 5-, and 10-coefficient multiplier schemes, respectively. Further, the error-reduction coefficient for each group of sub-intervals is derived by following the mathematical approach detailed in [HSa+20]. The binary representation of the proposed coefficients for multiplier and divider are also shown in Table 4.1.

The partitioning for each approach (e.g., 3-, 5-, or 10-coefficient) is implemented through a small-sized multiplexer (coded via casex statement in Hardware Description Language (HDL)). The LUT usage for such multiplexer-based implementation depends on: First, *the number of error-coefficients* as the *MUX-inputs* and second, *the complexity of conditional statements in the casex statement*. In order to minimize the former (i.e., number of MUX inputs), three schemes have been proposed, each with at most 10 coefficients. In order to minimize the latter (i.e., complexity of conditional statements for selecting the coefficients), only 4 MSBs of the fractional parts are considered for the partitioning. Moreover, in this partitioning, neighboring sub-regions having the same coefficient are packed to reduce the complexity for each conditional statement.

From the hardware-usage perspective, each 6-LUT functions as a 4-MUX (in which four inputs and the two select lines are the LUT inputs). Further, as also denoted by [Cha13], a 16:1 multiplexer requires a single FPGA-slice having four 6-LUTs. A squarish partitioning was also chosen through a MUX-based approach to minimize the overhead of the proposed error-reduction strategy. This is while checking the rhombus borders (see the original error shape in Figs. 4.4 and 4.5) is very costly in hardware and could nullify the LUT-saving, gained from the approximating approach.

It should be noted that the proposed partitioning is also scalable, as the resource cost for choosing one of a few coefficients does not grow exponentially, contrary to the REALM approaches. In fact, the proposed partitioning surpasses the state of the arts in terms of resource-error trade-off: Not only with 10 error-coefficients and considering four MSBs, the ARE of 0.6% was achieved, which is better than REALM (0.8% ARE, considering 3 MSBs), but also the resource footprint of this scheme is 193 LUTs, still lower than REALM counterpart (see Sect. 4.6.3).

Table 4.1 Binary representation of error-reduction coefficients in 16-bit multiplier and divider (3/4 MSBs are zero for MUL/DIV and excluded)

Multiplier			
3-Coefficient	5-Coefficient	10-Coefficient	
(1) 100000100111	(1) 1001111111111	(1) 1001111000110	(6) 0101110011111
(2) 010011101100	(2) 1000011011101	(2) 1000110110001	(7) 0100101000011
(3) 000100101001	(3) 0110010001010	(3) 0111111000100	(8) 0100001011101
	(4) 0011110010111	(4) 0111000110101	(9) 0011110000011
	(5) 0000111110000	(5) 0110010100011	(10) 0010101111111
Divider			
3-Coefficient	5-Coefficient	9-Coefficient	
(1) 1000011111111	(1) 1001111000100	(1) 1001110001111	(6) 0110010100101
(2) 0100010111111	(2) 1000001000111	(2) 1000110111100	(7) 0101000101011
(3) 0001011111111	(3) 0110110001101	(3) 1000000010100	(8) 0100111101000
	(4) 0101010100111	(4) 0111001100010	(9) 0100001101100
	(5) 0011011100100	(5) 0110100001101	

4.3.2 *Structure of RAPID Multiplier and Divider (Non-pipelined)*

Figure 4.6 illustrates the RAPID multiplier and divider structure discussed herein.

Leading-one detection To accelerate leading one detection in our FPGA-customized method, this process is calculated based upon 4-bit LODs (implemented by directly configuring LUTs). In the first step, the presence of bit value '1' is probed simultaneously in each 4-bit segment of the operands. To this end, one LUT is configured as a logical OR function, applied on 4-bits of each group to reveal whether the segment contains a bit with value '1' (acts as a zero-detection flag). In parallel, another 6-LUT is configured to two 5-LUTs in such a way that it determines the position of leading-one in the 4-bit segment (LOD4-LUT). Finally, based on the resulting bits from these 6-LUTs outputs, the position of the leading one in the most significant group is determined through the priority logic. For example, the position of leading one in the 8-bit LOD equals to the concatenation of {Location index of most significant segment, Leading one position in that segment}, e.g., the leading one position in "01010101" is {{1},{10} => {110}} in binary. The first part ({1}) is the output of Flag-LUT that has been employed for the upper 4-bit segment. The second part ({10}) is the result of LOD4-LUT on the upper segment. A similar method has also been exploited for 16- and 32-bit LODs. For example, in a 16-LOD, if the upper half of the operand is zero, the 16-bit LOD is equal to the lower 8-bit LOD. Else, the position of leading-one is 8+leading-one position in the upper-half 8-LOD. This can be obtained by applying a logical-OR function on the outputs of Flag-LUTs on the 4-bit segments of the upper half. In LeAp [ZSA20], LOD step was orches-

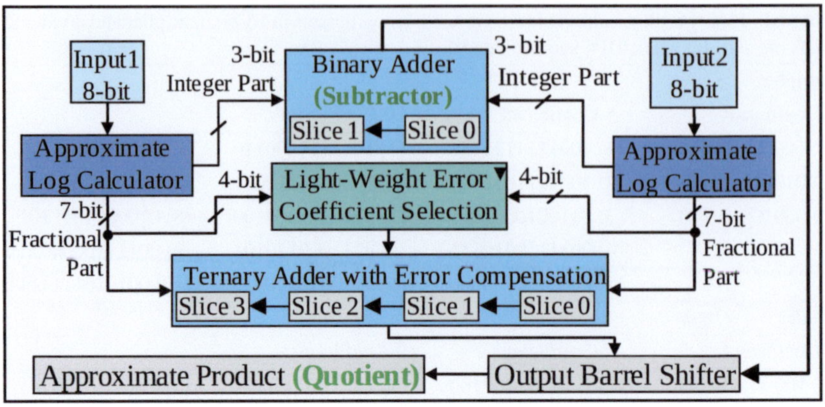

Fig. 4.6 Overall structure of proposed RAPID multiplier and divider

trated through an Finite-State Machine (FSM) and performed in, at most, five clock cycles. In order to achieve efficient pipelining and minimize the number of registers, the LOD is implemented as combinational logic. Subsequently, the critical path is analyzed to achieve balanced pipelining partitioning.

Addition of Integer Parts As shown in Fig. 4.6, each 4-bit addition is fulfilled by one Virtex-7 slice which includes four 6-LUTs and its associated fast carry chains. Together, these implement a CLA. Extending the 4-bit addition to 8-bits is easily achievable by connecting the C_{out} from a previous slice to the C_{in} of the next slice. Recalling Eqs. 2.6 and 2.7, the division is performed by changing additions to subtractions through 2's complement modules.

LUT-Optimized Ternary Addition Recalling that in LeAp [ZSA20], error-reduction coefficients were proposed such that they depend only on the fractional bits (and not on the intermediate result of Mitchell's MUL/DIV, contrary to MBM/INZeD [HHS18a, HHS19]). On the other hand, LUTs and their associated fast carry chain in Xilinx UNISIM library [Xil13] can be configured to implement a *ternary* adder. To this end, the LUTs and carry chain primitives of the FPGA were manually configured to implement a ternary adder. On the contrary, in REALM [HSa+20], MBM [HHS18a], and INZeD [HHS19], an additional circuit is needed to add the error-reduction term (or half of it) to the original Mitchell's circuits, based on the intermediate summation/subtraction of fractional parts. Such an approach increases the latency of these counterpart designs.

To prevent overflow in 2N-by-N bit standard division, the condition for dividend $< 2^N \times$ divisor needs to be satisfied [Par10]. This means that the scaling of the divider lies in the range of 2^0 to 2^{N-1} (similar to [HHS19]). Therefore, only N-1 bits are used from the subtractors for the output, and N LSBs from $\widetilde{\log}_{\text{dividend}}$ is neglected. This also reduces the resource footprint for the subtractor and barrel shifter and does not affect the accuracy.

4.3.3 Pipelined Architecture of RAPID Multiplier and Divider

In order to minimize the latency overhead, the combinational datapath of the multiplier and divider should be partitioned with a near-uniform latency over the pipelined stages. To achieve this, the following steps were adopted. First, each stage of the multiplication or division is synthesized in isolation to get an *estimation* of the delay for each stage. Figure 4.7 shows the latency of individual stages in non-pipelined and various pipelined configurations for a 5-coefficient multiplier and divider (the results for all designs are shown in Sect. 4.6.3). Note that after integrating the sub-modules and synthesizing the entire component, the end-to-end latency may change due to the tool's default optimizations and structural modification. Based on this analysis, the pipelining registers were inserted in the proper locations of the non-pipelined design. Afterward, re-synthesis was performed to assess whether a marginal fine-tuning for the adopted partitioning can result in better end-to-end latency.

It can be inferred from Fig. 4.7, that through the adopted pipelining approach, the latency of stages become similar. Moreover, the delay overhead of a pipeline register is small, particularly since the divider is in the critical path in most applications. The RAPID architecture supports flexible pipelining, allowing multipliers and dividers to be implemented with a user-selectable number of stages: each proposed soft-core design operates at a unique frequency, supporting deployment across different frequency domains of a single FPGA or various FPGA families. This configurability spectrum enables application-specific optimization of power-throughput and latency-throughput trade-offs. It is worth underlining that featuring fine-grained pipelining at an intra-unit granularity could enable operating at a higher frequency level versus coarser granularity (i.e., inter-unit), especially in applications that include division operation. Please note that although our circuits are specialized for FPGAs, our fine-grain pipelining approach is also applicable to ASIC-based architectures.

4.4 SIMDive: Approximate SISD/SIMD Soft Multiplier-Divider (Hybrid) with Tunable Accuracy

The ever-increasing quest for data-level parallelism and variable precision in ubiquitous multimedia, healthcare, and various NN-based applications has motivated using SIMD architectures. To address this call, SIMDive [EUK20] was designed as the first hybrid multiplier-divider with tunable accuracy, further customized to support a resource-efficient SIMD execution mode in FPGAs. In fact, SIMDive has enabled the plasticity to switch between the hybrid functionality of multiplier-divider on-the-fly, without the need for reconfiguration. To achieve this, the proposed LUT-based hybrid architecture is implemented based on Mitchell's algorithms, which transform multiplication (division) into addition (subtraction) in logarithmic representation. SIMDive is customized for FPGAs from two perspectives: (i) A novel and LUT-

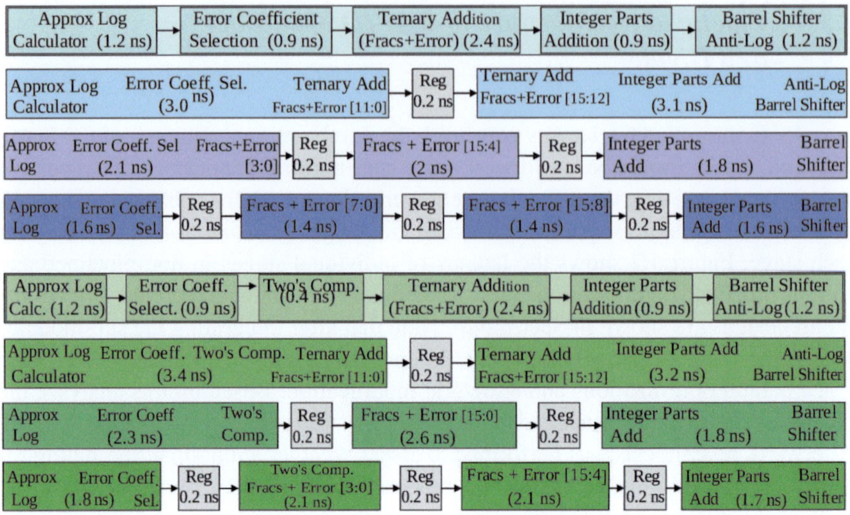

Multiplier	Latency	Relative Throughput	LUT	Total Reg
Non - Pipelined	6.1	0.8	171	64
2-stage Pipeline	6.6	1.5	173	98
3-stage Pipeline	6.9	2.12	174	126
4-stage Pipeline	7.2	2.71	175	141

Divider	Latency	Relative Throughput	LUT	Total Reg
Non - Pipelined	6.4	2.84	112	41
2-stage Pipeline	7.2	5.19	121	70
3-stage Pipeline	8.4	6.67	127	98
4-stage Pipeline	9.2	8.12	130	119

Latency = # Pipeline Stages × (Max of Stages' delay+Reg delay) $\text{Rel. Tput} = N \times \frac{\text{Latency of Non-pipelined Acc Unit}}{\text{Latency of pipelined Approx. Unit}}$

Fig. 4.7 Proposed non-pipelined and 2-, 3-, and 4-stage pipelined structure of 16×16 RAPID-5 multiplier (blue) and RAPID-9 16/8 divider (green)

specialized approach for implementing error-reduction circuit and (ii) the addition of error-coefficients through a ternary adder. The hybrid multiplier-divider supports precision variability in a SIMD mode, e.g., from 32- to 16-, and 8-bit. Therefore, it is an appropriate candidate for dynamic precision scaling techniques and various applications featuring SIMD potentials, e.g., NNs. In the following, the error-reduction schemes of SIMDive are presented, followed by an elaboration on the structure of this hybrid unit, which supports the SIMD execution mode.

4.4.1 SIMDive Light-Weight and Tunable Error-Reduction Scheme

Mitchell's error for different sizes of multiplier and divider is shown in Fig. 4.8 (based on Eqs. 4.3 and 4.4 shown in Sect. 4.3). Figure 4.8 exhibits the non-uniform but symmetrical error distribution in both multiplier and divider: Errors tend to be the same at the beginning and end of each power-of-two interval, encouraging the same reduction approach for all multiplier or divider sizes.

In the SIMDive approach, the squarish region for all combinations of inputs is subdivided to $2^3 \times 2^3 = 64$ sub-regions by merely using 3 MSBs of inputs. Hence, a distinctive coefficient is assigned to each region, representing the average error for each sub-interval. To derive the error-reduction coefficients that minimize the average relative error in each sub-region, a mathematical approach was followed.[3]

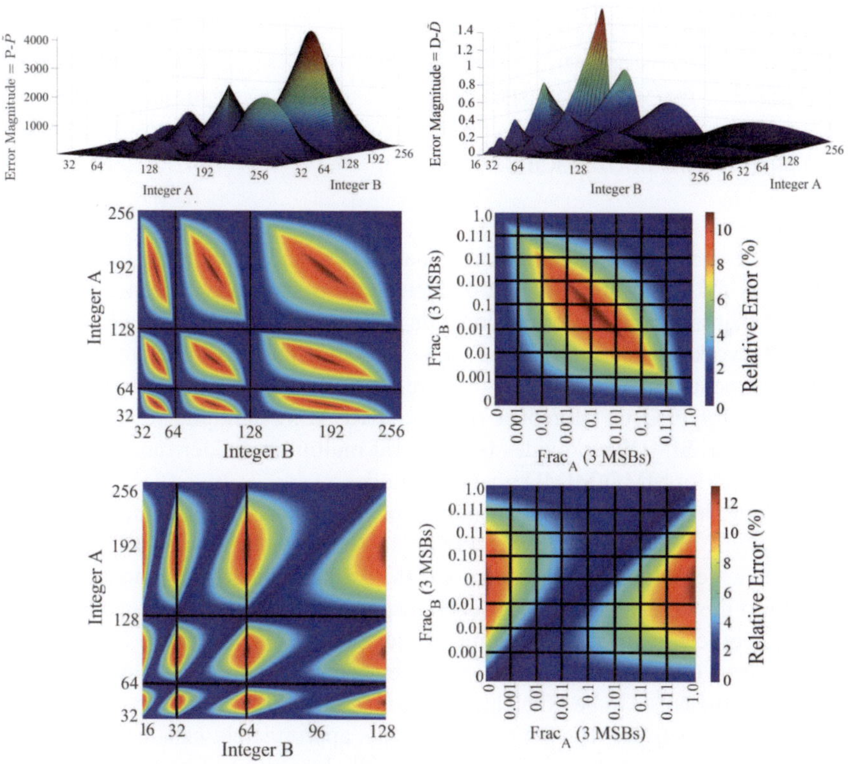

Fig. 4.8 Mitchell's error: 8 × 8 multiplier (top-left), 16/8 divider (top-right). Middle (bottom) rows show error scaling with multiplier (divider) size. Right panels illustrate SIMDive's error-reduction: the fractional zone is divided into 64 regions, each with a specific coefficient

[3] The mathematical formulations are detailed in [HSa+20].

Table 4.2 Configuration INIT value (hexadecimal) for error-reduction LUTs in SIMDive multiplier and divider

Multiplier			
Coeff [15]	0000000000000000	Coeff [7]	7AC120E9148D9AD2
Coeff [14]	0000000000000000	Coeff [6]	43012121852990E8
Coeff [13]	0000000000000000	Coeff [5]	D592D1A2C0A5336B
Coeff [12]	0000000C18100000	Coeff [4]	94190083448C1259
Coeff [11]	00061E32266C7800	Coeff [3]	0018123B00013397
Coeff [10]	00396D566C2A2460	Coeff [2]	0102021082046913
Coeff [9]	0F4A2401C5B6F69C	Coeff [1]	1551008100820650
Coeff [8]	361C90D046999F12	Coeff [0]	9500000100010103
Divider			
Coeff [15]	0000000000000000	Coeff [7]	04AD3A5428BA3D06
Coeff [14]	0000000000000000	Coeff [6]	D4D18E4EA73F3994
Coeff [13]	0000000000000000	Coeff [5]	F6661C4A32DB3527
Coeff [12]	0000018181810000	Coeff [4]	4243DC139C9718A2
Coeff [11]	000386466262E100	Coeff [3]	542DDC1C0C97AC12
Coeff [10]	008C482AD6D0D8E0	Coeff [2]	22B2367E87019850
Coeff [9]	0E90D2614478D498	Coeff [1]	DA48B05937950D5A
Coeff [8]	B375AAD2693CB34C	Coeff [0]	B8427A22AF76B33F

For the sake of reproducible results, the final INIT hexadecimal value for the 16-bit coefficients for both multiplier and divider are also shown in Table 4.2. It should be noted that although the error-reduction approaches of SIMDive and REALM [HSa+20] works are similar in spirit, our work has pursued a different architectural implementation, i.e., proposing a novel LUT-based implementation approach, specifically for FPGAs. Moreover, while the REALM multiplier operates solely in SISD mode, SIMDive provides both hybrid multiplier-divider functionality and native support for SIMD/MIMD operation modes.

To achieve a lightweight implementation of the proposed error-reduction scheme and to benefit from the underlying FPGA structure, the following approach was proposed, which efficiently utilizes 6-LUTs: 3 MSBs of each fractional part are assigned to the inputs of the LUTs, each responsible for calculating one bit of the 64 error coefficients. In fact, 64 output entries of ith LUT determine ith bit of the 64 coefficients (in binary representation). Therefore, using only 8, 16, and 32 LUTs, 64 error coefficients can be efficiently determined for the target 8-, 16-, and 32-bit operations, respectively. Having 64 coefficients appropriately calculated based on the combination of both operands addresses both drawbacks of previous studies in the literature, i.e., neglecting the magnitude of error due to separately approximating each operand and overflow cases (in fact, the overhead of separately handling such overflow cases is not clearly discussed in MBM [HHS18a] and INZeD [HHS19].) Our approach is also highly suited for LUT-oriented platforms, as it utilizes the LUTs and their associated fast carry chain in Xilinx UNISIM library [Xil13] to implement a ternary adder.

Applicability of SIMDive error-reduction scheme to other FPGAs: Our approach is easily applicable to various FPGAs without the need for architectural modification. It can achieve even better accuracy if the target LUT has more than six inputs (e.g., 8-bit ALMs in Intel's Stratix and Arria series), considering 4-bits of each fractional part will provide 256 sub-regions. Therefore, solely one 8-LUT can enable 256 error-reduction coefficients, which can significantly improve accuracy.[4] Our error analysis reveals that by using 8-LUTs, the proposed light-weight error-reduction approach in this work can significantly reduce and bound the ARE to $< 0.1\%$.

4.5 The Structure of SIMDive SIMD Multiplier-Divider

The overall structure of the proposed SIMD multiplier-divider is illustrated in Fig. 4.9. The controlling signals, i.e., *precision* and *Mul/Div mode* shown in Fig. 4.9a, serve to establish diverse sub-word size and functionalities of each module, respectively. The default one-hot encoding preferred by FPGA manufacturers was used, as it has been shown to be more resource-efficient than binary encoding for reconfigurable fabrics [Ash10]. One-hot encoding scheme has two advantages. First, supporting *mixed-precision* and *mixed-functionality* for the first time: The proposed SIMD multiplier-divider can either operate as a single 32×32 unit or be decomposed into a twin 16×16, one 16×16 and two 8×8, or quad 8×8 units each of which can act separately as multiplier or divider. Supporting dynamic mixed-functionality in our design eliminates the need for separate resources and the overhead of reconfiguration and accommodates both operations in a single module. This feature is highly desirable, especially in multiplication-intensive workloads with fewer divisions (appealing for NN and multimedia). Moreover, in the case of sub-word parallel processing, a complete 32-bit SIMDive unit is not occupied for a division, and it can also be configured to multiplier simultaneously, if needed. Second, the separate *data-size* signals can also be used to power-gate each sub-unit.

Referring to Fig. 4.9b, the detection of leading one in our proposed approximate log calculator is orchestrated in parallel with extracting the fractional part (for each 4-bit segment of the input), by using two 6-LUTs: The first 6-LUT acts as a zero-detection flag and detects whether the four bits are zero. The second 6-LUT (used as two 5-LUTs) is directly configured in such a way as to reveal the position of leading-one (0–3). Afterward, depending on the required operand word length, integer and fractional parts are determined based on the most significant non-zero segment. Based on our analysis, 4-bit segmentation renders the best energy-delay product for 32-bit SIMD architecture. Afterward, as shown in Fig. 4.9b, each 4-bit addition—for adding the integer parts or adding the fractional parts—are fulfilled by a Virtex-7 slice. As shown in part (c) of this figure, each slice includes four 6-LUTs and its

[4] Ternary addition also can be implemented efficiently in Stratix and Arria series.

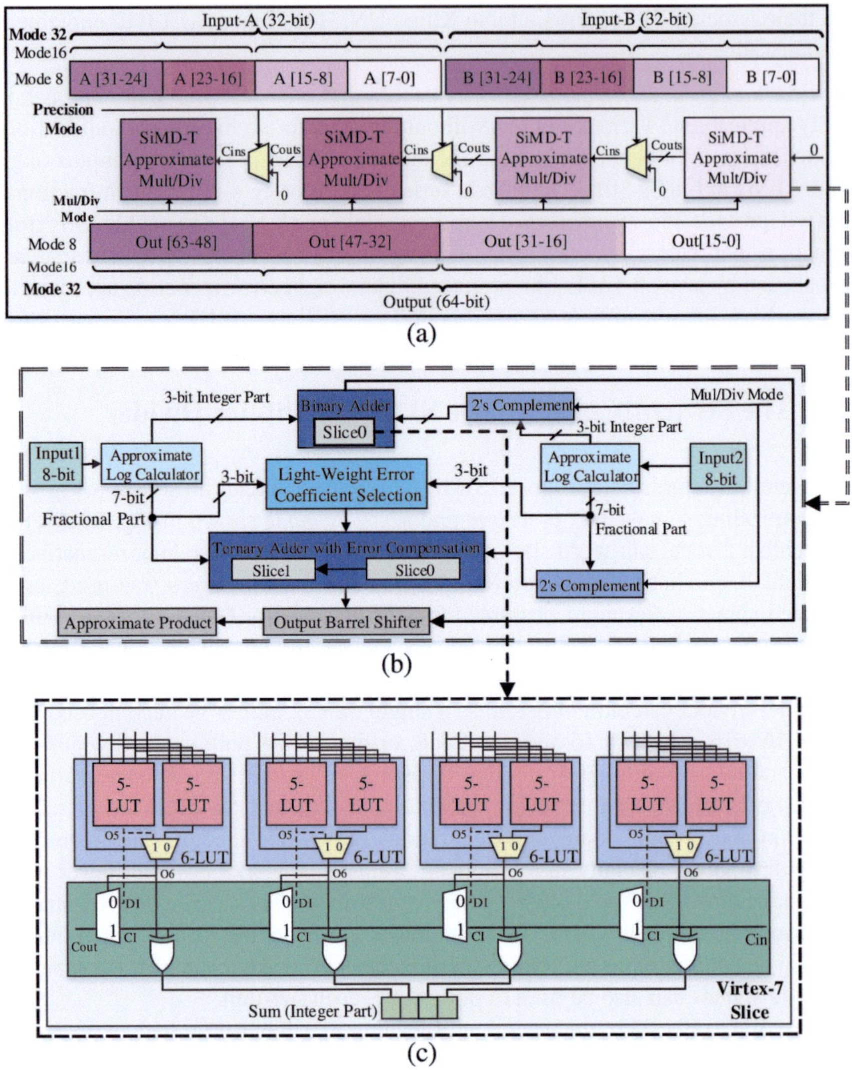

Fig. 4.9 **a** SIMDive structure, **b** proposed 8-bit multiplier/divider, **c** Virtex-7 slice (used for addition/subtraction of integer/fractional parts)

associated fast carry chains, together implementing a CLA adder. Extending the 8-bit addition to 16- and 32-bits in our SIMD architecture is also easily achieved by connecting the C_{out} from the previous adder to the C_{in} of the next adder, handled by yellow multiplexers in part (a) of this figure. The division is also performed by altering additions to subtractions. The subtraction operation is implemented with a 2's complement module, which calculates the negative of the second input before feeding it to the adder.

Shared modules in hybrid/SIMD modes: besides the two's complement modules, integrating the multiplier and divider into a hybrid unit marginally affects the complexity of the circuit by including 2-MUX units to select the mode in sub-modules (except LOD/final barrel shifter that is equal in both modes). On the other hand, in order to support simultaneous scalings for sub-word length in SIMD mode, the LOD, error-reduction, and final shifter become more complex. For example, in 32-bit LOD, instead of 5 for SISD mode, $4 \times 3 = 12$ bit is used to also support simultaneous 8-bit leading one detection in SIMD mode (this is similar for integer part adder). On the other hand, the complexity of the larger adder that is employed for fractional parts has not increased significantly, as the carry-out of the consequent 8-bit is used as carry-in in the following 8-bit for 16- and 32-bit modes. It is worth noting that there is also a small similarity between SIMDive error coefficients, as 3 and 4 MSBs are similar in each multiplier and divider. In short, although the overall overhead of additional circuitry for managing hybrid and SIMD modes is not negligible in 8-bit hybrid design, it gets dwarfed in larger SIMD designs (see Sect. 4.6.4), thanks to transforming the 2D structure of array multiplier (divider) to 1D adder (subtractor) in the logarithmic representation.

SIMDive is highly suitable to be utilized in a cross-layer approximation methodology (which will be presented in Sect. 7.4.1), and its rationale is multi-fold. First, Such resource-efficient SIMD architecture bodes well for a cross-layer approach as it enables a chain of techniques, i.e., precision scaling on top of Mitchell-based multiplier-divider (detailed results are presented in Sect. 4.6.4). Especially in the case of division, SIMDive reduces the drastically high latency of an accurate divider, nearly to the latency of a same-size multiplier. Second, it yields high accuracy, especially negligible error bias, which, as mentioned earlier, can play a pivotal role in an approximation of consecutive kernels having an aggregation-based structure. Third, the plasticity provided by SIMDive hybrid mode enables on-the-fly switching between two functionalities without the need for reconfiguration.

4.6 Circuit-Level Results and Discussion

4.6.1 Experimental Setup and Tool-Flow

Implementation and Hardware Performance Metrics We have evaluated LeAp, SIMDive, and RAPID multipliers and dividers against accurate and multiple cutting-edge approximate multipliers and dividers in the literature.[5] The designs are developed in Verilog/HDL, synthesized, and implemented in Xilinx Vivado on Virtex-7 FPGA. To ensure the scalability of multipliers and dividers, they are implemented and compared for comprehensive precision of 8-, 16-, and 32-bit. The area and latency

[5] The results of the circuit-level works are presented based on their publication date (i.e., LeAp $\rightarrow$ SIMDive $\rightarrow$ RAPID) and comprehensiveness of the simulations.

Table 4.3 Summary of design metrics in 16 × 16 LeAp and baseline multipliers

Approach	Area (6-LUT)	Throughput (/µS)	Power (mW)	Energy (µJ)	ARE (%)	PRE (%)
Accurate [XilinxIpCore]	280	203	31.5	328	–	–
CA [Ull+18]	245	201	30.4	341	0.3	19.04
CC [Ull+18]	240	416	32.7	306	14.88	96.7
Trunc12 × 12	139	221	24.1	242	0.2	100
Mitchell [NMi62] comb.	249	212	27.1	276	3.85	11.11
Mitchell [NMi62] seq.	**121**	**138**	**15.4**	264	3.85	11.11
MBM [HHS18a]	128	94	17.5	331	2.63	8.81
LeAp-2	132	238	21.6	**224**	1.56	6.67
LeAp-3	135	224	23	232	1.23	5.72
LeAp-5	139	217	24.7	246	**0.98**	**4.76**

are directly collected from Vivado reports. Power and energy dissipations[6] are scrutinized through simulations in Xilinx Power Estimator (XPE) over 100 million inputs, uniformly distributed in a random order over the whole input interval.

Accuracy Metrics To calculate the accuracy metrics, i.e., ARE, PRE, and EB, the behavioral structure of multipliers and dividers are developed in C++. For 8- and 16-bit designs, an exhaustive test is applied. For 32-bit error characterization, 2^{32} pair of inputs (which is about 4.3 billion pairs), have been evaluated in Monte Carlo simulations.[7] These inputs are randomly, yet uniformly distributed over the whole 32-bit interval.

4.6.2 Results for LeAp Multipliers

Design metrics and error analysis are summarized in Tables 4.3 and 4.4, where x in "LeAp-x," denotes the number of error-reduction coefficient. The following conclusions are notable based on the results:

LeAp Versus PP-Based Multiplication: Comparing Mitchell- with truncated- and hierarchical-based multipliers, designed based on incorporating smaller inexact instances [Ull+18] justifies three points: (1) Thanks to the translation of multiplication to addition, area of Mitchell-based multipliers grows by factor of ~2.5 while

[6] Similar to prior works, only dynamic power is reported from Xilinx Vivado, as the static power analysis of Vivado is for the entire FPGA [Ull+21, SUl+20].

[7] Simulated on Rack Server with Intel Xeon E5-2667 CPU @ 3.20 GHz and 512 GB RAM.

Table 4.4 Summary of design metrics in 32 × 32 in LeAp and baseline multipliers

Approach	Area (LUT)	Throughput (/μS)	Power (mW)	Energy (/μJ)
Accurate [XilinxIpCore]	1103	163	65.8	672
Mitchell [NMi62]	**246**	53	**23.5**	490
MBM [HHS18a]	258	42	28.2	635
LeAp-2	334	**187**	38.1	**423**
LeAp-3	337	174	40.8	437
LeAp-5	342	166	43.3	451

compared to the area of other approaches which grows by a factor of $\sim$ 3.7. This further highlights the efficiency of LeAp variants, especially in larger bit-width. (2) Although the ARE of Trunc12 × 12 is better than 16 × 16 LeAp, referring to the previous point, in larger multipliers, more LSBs need to be truncated, i.e., Trunc18 × 18 have a same area to 32 × 32 LeAp. This will further deteriorate its accuracy and increase the cases with a large error, i.e., near or equal to 100% (hundreds of millions of cases), while in LeAp, most of the errors are <1%. (3) approximation applied on hierarchical multipliers is rewarding in accuracy-resource trade-off only when the analysis is done from scratch for each multiplier size. Otherwise, errors can be accumulated in larger designs. This will significantly sacrifice output accuracy to gain resource efficiency.

LeAp Corroborates Its Superiority by Improving Resource Consumption: Augmenting Mitchell's algorithm with our proposed log calculator architecture—efficiently customized for FPGAs—improves all the performance metrics compared to the accurate multiplier (LeAp variants enable various performance-accuracy trade-off levels). In contrast, both CA [Ull+18] and MBM [HHS18a] designs dissipate even more energy (with lower throughput) than an accurate multiplier. Interestingly, the resource gains of LeAp become even more pronounced in the 32×32 multiplier. In addition, increasing the number of coefficients boosts the precision with a negligible cost, thanks to the implementation approach (i.e., coefficient selection only depends on a few MSBs of fractional parts, and one more error-coefficient would only increase the conditional statements by one). Most of our design's extra LUTs were used to implement the proposed fast log calculator. This has improved the total execution time and contributed to reducing energy. Moreover, interestingly, LeAp has improved its energy levels compared to its baseline counterparts. In fact, although the error-reduction scheme of MBM has a smaller area and power than LeAp, its prolonged critical path and lack of a fast approximate log calculator have resulted in longer critical path delay (lower throughput) and, therefore, higher energy requirement.

The Error-Reduction Approach of LeAp: Compared to Mitchell's multiplier, both ARE and PRE error metrics are enhanced significantly in LeAp designs. Note that although an exhaustive 32 × 32 test is massively time-consuming, unlike hierarchical designs, Mitchell's ARE-error does not significantly change in larger multipliers. In fact, the proposed error-reduction scheme is independent of input size and outperforms the MBM design w.r.t both error metrics.

4.6.3 Results for RAPID Multipliers and Dividers

Design metrics and error analysis are summarized in Tables 4.5 and 4.6, where x in "RAPID-x," denotes the number of error-reduction coefficient. The designs are compared with respect to different performance metrics, including *Performance (throughput) per Watt*, which is regarded in the industry as the new Moore's Law [ARM21]. The following conclusions are notable based on the results:

RAPID Multipliers and Dividers Versus DSP-Based and Accurate Vivado IPs: the results corroborate the statement of prior studies [BYB18, SLe+19] that DSPs are able to be efficiently utilized only for large bit-width precision. The reason behind this degraded performance is that DSP48E1 hosts a 25×18 hard-wired multiplier and is not optimized for smaller multipliers. This has been the main motive that FPGAdesigners have recommended utilizing soft multipliers should be used for implementing lower bit-widths [BYB18, SLe+19, Xil18a, Xil16b]. In particular, compared to LUT-based implementation, DSP-based dividers are less energy-efficient for both 8- and 16-bit precision. Targeting higher-order precision of 32-bit, the proposed RAPID multiplier and divider have lower latency than DSP-based counterparts. In terms of energy efficiency (relative energy/instruction), DSPs are better than RAPID, only for multiplication. In contrast, RAPID dividers are significantly better than DSP-based implementation (especially in 32-bit precision). Compared to LUT-based IPs, the results exhibit the efficiency of the RAPID divider in all performance metrics when targeting division of any precision. In the case of multiplication, although accurate IP renders better performance in 8-bit over RAPID, the area- and/or energy-saving of RAPID over accurate Vivado IP becomes substantial in 16-bit. Moreover, the latency of RAPID also becomes smaller when targeting 32-bit. Overall, the improvements become more pronounced when architectures are implemented for higher bit-width. This is due to: first, transforming the 2D array-based structure of Mul/Div to 1D Add/Sub through Mitchell's algorithms, as discussed earlier. Second, the cost of error coefficients gets amortized in higher-order designs.

RAPID Multipliers and Dividers Render Better Performance over Hierarchical Structures: comparing Mitchell-based RAPID with modular counterpart AFM (structured by incorporating smaller inexact instances) demonstrates three points: First, thanks to transforming the 2D array structure of multiplication to 1D addition in the logarithmic representation, the area of Mitchell-based unit grows by the factor of $\sim$2.6, less aggressive compared to $\sim$3.7 for the array-based counterparts. This further accentuates the efficiency of RAPID, particularly in larger bit-width. Second, comparing the results of 8- versus 16-bit asserts that approximation applied on hierarchical approaches is beneficial in accuracy-resource trade-off only when it is done from scratch for each multiplier size; otherwise, accumulated error in larger designs

Table 4.5 Accuracy-resource trade-off of accurate/approximate multipliers (pipelined/non-pipelined) in an FPGA-based implementation

Multiplier

8 × 8 Mul	LUT	FF	E2E[a] Latency (ns)	Rel.[b] Tput	Circuit power (mW)	Clk power (mW)	Rel.[c] energy per inst	Rel. Tput/ Watt	ARE (%)	PRE (%)	Error Bias (%)
DSP-based	7 + 1 DSP	**31**	3.72	1.97	17.52	**10.3**	1.77	**1.13**	–	–	–
Acc IP_NP[4]	60	48	**3.67**	1.0	15.91	–	**1.0**	1.0	–	–	–
Acc IP_P2	60	88	4.30	1.71	23.19	37.5	2.23	0.45	–	–	–
Acc IP_P3	60	112	5.10	**2.16**	47.5	43	2.64	0.38	–	–	–
RAPID-3_NP	57	39	4.97	0.74	**12.06**	–	1.03	0.97	1.02	6.1	0.06
RAPID-5_P2[5]	62	56	5.45	1.35	18.17	22.9	1.91	0.52	0.91	**4.45**	**0.05**
RAPID-10_P3	71	69	6.88	1.60	24.31	34.0	2.29	0.44	**0.64**	3.69	0.05
AFM1	69	58	5.50	0.67	18.66	–	1.76	0.57	0.23	16.52	0.23
SIMDive-MUL	61	32	5.13	0.72	16.34	–	1.44	0.7	0.82	4.76	0.05
MBM	64	33	5.16	0.71	17.48	–	1.54	0.65	2.60	8.59	0.09
Mitchell	**51**	32	4.82	0.76	13.39	–	1.11	0.9	3.77	11.11	3.77
DRUM-4	**53**	32	5.08	0.72	14.33	–	1.25	0.81	5.82	25.35	1.84
16×16 Mul	LUT	FF	E2E latency (ns)	Rel. Tput	Circuit power (mW)	Clk power (mW)	Rel. energy per inst	Rel. Tput/ Watt	ARE (%)	PRE (%)	Error bias (%)
DSP-based	8 + 1 DSP	**32**	**4.11**	1.19	**17.48**	**9.8**	**0.48**	**2.08**	–	–	–
Acc IP_NP	287	64	4.88	1.0	47.81	–	1.0	1.0	–	–	–
Acc IP_P2	249	176	6.14	1.60	64.86	65.3	1.71	0.59	–	–	–
Acc IP_P3	249	245	8.88	1.65	94.49	75.0	2.15	0.46	–	–	–
Acc IP_P4	249	343	9.60	2.03	150.73	130.0	2.89	0.35	–	–	–
RAPID-3_NP	**168**	64	5.90	0.83	31.43	–	0.86	1.17	1.03	6.1	0.06
RAPID-3_P2	169	98	6.11	1.52	47.35	40.7	1.21	0.85	1.03	6.1	0.06
RAPID-5_P3	177	126	6.87	2.25	75.35	58.2	1.27	0.81	0.93	**4.45**	0.05
RAPID-10_P4	193	141	7.25	**2.52**	84.75	87.4	1.46	0.7	**0.56**	3.69	0.23
AFM1	261	66	7.32	0.67	44.78	–	1.41	0.71	1.34	17.80	1.34
SIMDive-MUL	216	64	5.95	0.82	37.06	–	0.95	1.06	0.82	4.90	0.05
MBM	204	65	6.59	0.74	35.34	–	1.0	1.0	2.63	8.83	0.09
Mitchell	**167**	64	5.51	0.99	31.46	–	0.64	1.56	3.85	11.11	3.85
DRUM-6	233	64	5.34	0.91	38.43	–	0.88	1.14	1.47	6.31	**0.04**

(continued)

Table 4.5 (continued)

Multiplier

32×32 Mul	LUT	FF	E2E Latency (ns)	Rel. Tput	Circuit Power (mW)	Clk Power (mW)	Rel. Energy per inst	Rel. Tput/ Watt	ARE (%)	PRE (%)	Error Bias (%)
DSP-based	18 + 4 DSPs	**75**	6.93	1.04	**35.48**	**14.8**	**0.47**	**2.28**	–	–	–
Acc IP_NP	1012	128	6.69	1.0	110.56	9-	1.0	1.0	-9	-9	-9
Acc IP_P2	1012	291	7.82	1.71	154.26	92.8	1.31	0.76	-9	9-	-9
Acc IP_P3	1012	835	9.74	2.06	218.32	146.6	1.6	0.63	-9	-9	9-
Acc IP_P4	1014	1119	11.73	2.28	330.64	250.8	2.31	0.43	-9	-9	9-
RAPID-3_NP	**434**	128	**6.30**	1.06	90.08	–	0.77	1.3	1.05	6.1	0.07
RAPID-3_P2	450	184	6.74	1.98	131.22	62.7	0.88	1.13	1.05	6.1	0.07
RAPID-5_P3	462	242	7.38	2.72	181.53	95.1	0.92	1.09	0.95	**4.45**	**0.05**
RAPID-10_P4	490	276	8.04	**3.33**	253.12	124.2	1.03	0.97	**0.58**	3.64	0.06
AFM1	995	317	10.76	0.62	119.63	–	1.74	0.57	2.88	22.40	2.88
SIMDive-MUL	521	128	6.88	0.97	83.50	–	0.78	1.29	0.91	4.72	0.05
MBM	533	129	7.51	0.89	89.94	–	0.91	1.09	2.69	8.74	0.10
Mitchell	**428**	128	**6.23**	1.12	60.25	–	0.55	1.83	3.91	11.11	3.91
DRUM-6	616	128	6.35	1.05	88.85	–	0.76	1.31	1.53	5.88	0.05

[a] End-to-end. [b] Throughput. [c] Energy = dynamic power × clock period. [d] Non-pipelined [e] 5-Coeff. 2-stage pipelined

significantly sacrifices output accuracy to achieve resource efficiency. On the other hand, accuracy metrics in Mitchell-based designs do not undergo notable changes. Finally, as already discussed in earlier work [ZSA20], some modular architectures sacrifice delay for LUT saving [Ull+18].

RAPID Designs Establish Better Resource-Accuracy Trade-Off Than Leading-One Based Truncated Counterparts: The higher accuracy levels of DRUM and AAXD come at the cost of higher resource consumption than logarithmic counterparts (due to using an *accurate* core unit). Furthermore, employing an accurate instance of a divider still results in a long latency multiple times that of a same-size multiplier. RAPID, on the other hand, has reduced the high latency of the accurate divider nearly to the latency of its same-size multiplier. Overall, our proposed architectures yield better resource-accuracy trade-offs, especially in higher bit-width. Finally, there are many cases with an error near or equal to 100% in the truncation-based AAXD divider. Such high error cases can result in false positive peaks in heartbeat and corner detection, as will be discussed in Sect. 6.4.

The Proposed Error-Reduction Outperforms Other Mitchell-Based Architectures: RAPID's error-reduction strategy surpasses that of REALM in three aspects. First, the proposed error-reduction scheme achieves a higher accuracy level, even with fewer coefficients/LUTs. This is due to the better partitioning scheme of RAPID,

Table 4.6 Accuracy-resource trade-off of accurate/approximate dividers (pipelined/non-pipelined) in an FPGA-based implementation

Divider

8/4 div	LUT	FF	E2E latency (ns)	Rel. Tput	Circuit power (mW)	Clk power (mW)	Rel. energy per inst	Rel. Tput/ Watt	ARE (%)	PRE (%)	Error bias (%)
DSP-based	11+ 2 DSPs	63	7.48	1.44	19.80	**11.6**	2.25	0.44	–	–	–
Acc IP_NP	51	42	10.74	1.0	9.62	–	1.0	1.0	–	–	–
Acc IP_P2	55	59	13	1.67	21.09	18.4	2.43	0.41	–	–	–
Acc IP_P4	69	89	23.7	1.81	26.27	29.1	3.14	0.32	–	–	–
RAPID-3_NP	41	**20**	**5.20**	2.06	9.72	–	0.48	2.06	0.99	5.74	0.02
RAPID-5_P2	44	20	5.18	4.15	14.74	20.8	0.88	1.13	0.79	4.34	**0.01**
RAPID-9_P3	51	20	5.34	**8.05**	22.66	30.6	0.91	1.47	**0.58**	**3.48**	0.01
SIMDive-DIV	44	20	5.23	2.09	9.23	–	**0.45**	**2.20**	0.77	5.20	0.01
INZeD	47	21	6.12	1.75	11.14	–	0.65	1.53	2.93	9.53	0.02
Mitchell	**36**	20	**5.10**	2.11	**8.53**	–	**0.42**	**2.40**	3.90	13.00	3.90
SAADI-EC (16)	103	42	13	0.84	24.42	–	2.99	0.33	2.37	8.82	1.92
AAXD (6/3)	**38**	20	6.06	1.77	**9.01**	–	0.52	1.91	2.08	100.00	1.49
16/8 div	LUT	FF	E2E latency (ns)	Rel. Tput	Circuit Power (mW)	Clk power (mW)	Rel. energy per inst	Rel. Tput/ Watt	ARE (%)	PRE (%)	Error bias (%)
DSP-based	197+ 7DSP	131	10.36	1.76	51.62	25.1	2.34	0.43	–	–	–
Acc IP_NP	169	76	18.23	1.0	**17.97**	–	1.0	1.0	–	–	–
Acc IP_P2	175	104	19.59	1.86	38.82	**12.3**	1.47	0.68	–	–	–
Acc IP_P4	181	168	20.09	3.63	56.21	24.5	1.68	0.6	–	–	–
RAPID-3_NP	**112**	41	6.38	2.98	18.67	–	**0.34**	**2.98**	1.02	5.74	0.02
RAPID-5_P2	121	70	7.07	5.16	27.77	19.5	0.49	2.04	0.79	4.34	0.01
RAPID-9_P3	127	98	8.35	6.62	30.29	24.3	0.44	2.28	**0.58**	**3.48**	**0.01**
RAPID-9_P4	130	119	9.2	**8.01**	34.68	27.7	0.42	2.40	0.58	3.48	0.01
SIMDive-DIV	143	64	**5.68**	3.28	23.84	–	0.39	2.57	0.78	5.20	0.01
INZeD	165	**41**	6.28	2.90	27.50	–	0.51	1.97	2.93	9.54	0.02
Mitchell	**106**	64	**5.56**	3.39	**17.34**	–	**0.32**	**3.11**	4.11	13.00	4.11
SAADI-EC (16)	342	126	25.70	0.71	57.01	–	4.31	0.23	2.14	8.82	1.76
AAXD (8/4)	151	155	12.51	1.46	25.17	–	0.93	1.08	2.99	100	0.90
AAXD (12/6)	207	233	21.26	0.86	34.51	–	2.16	0.46	0.74	100	0.30

(continued)

Table 4.6 (continued)

Divider

32/16 Div	LUT	FF	E2E latency (ns)	Rel. Tput	Circuit power (mW)	Clk power (mW)	Rel. energy per inst	Rel. Tput/ Watt	ARE (%)	PRE (%)	Error bias (%)
DSP-based	11+ 2 DSPs	63	7.48	1.44	19.80	**11.6**	2.25	0.44	–	–	–
DSP-based	359+ 9DSP	228	11.65	3.62	92.91	28.7	0.98	1.02	–	–	–
Acc IP_NP	597	139	42.24	1.0	**34.16**	–	1.0	1.0	–	–	–
Acc IP_P2	607	208	47.48	1.78	74.55	**10.4**	1.4	0.74	–	–	–
Acc IP_P4	607	339	56.95	2.97	111.04	21.9	1.75	0.76	–	–	–
RAPID-3_NP	378	85	**6.70**	6.30	42.70	–	**0.19**	5.04	1.04	5.74	0.02
RAPID-5_P2	389	128	8.43	10.02	49.07	16.5	0.2	5.22	0.79	4.34	**0.01**
RAPID-9_P3	399	173	9.86	12	59.20	20.2	0.2	5.24	**0.61**	**3.48**	0.01
RAPID-9_P4	417	213	11.10	**15.22**	67.59	26.7	0.2	**5.53**	0.61	3.48	0.01
SIMDive-DIV	381	**80**	6.84	6.18	43.04	–	0.2	4.90	0.81	5.16	0.02
INZeD	422	81	8.15	5.18	47.67	–	0.27	3.71	2.96	9.47	0.03
Mitchell	**349**	80	**6.21**	6.91	39.42	–	**0.17**	**5.99**	4.19	13.00	4.19
SAADI-EC (16)	822	228	51.60	0.82	92.86	–	3.32	0.30	2.33	9.04	1.85
AAXD (8/4)	**361**	278	24.66	1.71	40.78	–	0.70	1.43	3.04	100	1.10
AAXD (12/6)	513	505	37.20	1.14	57.95	–	1.49	0.67	0.79	100	0.35

[a] End-to-end. [b] Throughput. [c] Energy per instruction = dynamic power × clk period. [d] Non-pipelined [e] 5-Coeff. 2-stage pipelined

which still considers four MSBs of fractional parts. Second, the exponential increase in the number of coefficients (256 for four MSBs) of REALM not only poses a noticeable resource penalty but also nullifies the gain when realized in a LUT-based implementation. Overall, RAPID enables a more cost-effective strategy for partitioning the squarish zone. Third, in addition to the error-reduction strategy, RAPID can achieve a higher throughput, energy, and throughput per Watt compared to SIMDive in non-pipeline configuration (except for 8- and 16-bit division). However, as noted previously, the higher throughput enabled by pipelining RAPID comes at the cost of increased throughput per Watt when compared to SIMDive.

Herein some of the key differences between RAPID and SIMDive are discussed. Supporting higher throughput through SIMDive has added to the complexity of the sub-modules. For example, $4 \times 3 = 12$ bit is used for LOD in 32-bit SIMDive, or 2- and 4-MUX units are used to select the functionality and sub-word length in the intermediate adders. Moreover, the integration of a multiplier and divider into a hybrid design marginally affects the complexity of the circuit by including 2-MUX units to select the mode in sub-modules. Finally, in order to support simultaneous scalings for sub-word length in SIMD mode, the LOD, error-reduction, and final shifter become more complex. For example, in a 32-bit LOD, instead of 5-bit for the SISD mode,

$4 \times 3 = 12$ bit is used to also support simultaneous 8-bit leading one detection for the SIMD mode (the complexity overhead is also posed to integer/fractional part adder).

Comparing Latency, Throughput, and Throughput per Watt in Pipelined and Non-pipelined Designs: Overall, operating at a higher frequency and producing one-operation-per-cycle through pipelining has resulted in additional flip-flops, increased end-to-end latency, and higher dynamic power dissipation. However, pipelining enables significant improvement in throughput. In fact, higher throughput enabled by increasing the number of pipeline stages comes at the cost of lower *Throughput per Watt* as well. This descending trend is mainly due to the increase in the number of FFs and end-to-end latency in such designs. In particular, the analysis of latency and throughput per Watt highlights the efficacy of pipelining for RAPID circuits from three perspectives. First, comparing accurate- and RAPID-pipelined designs demonstrates that even with increasing the number of pipelining stages, the end-to-end latency for each x-stage-based RAPID remains smaller than its x-stage-based accurate counterpart (except for 8-bit multiplication). Second, the x-stage-based pipelined RAPID enables higher throughput per Watt versus its x-stage-based *accurate* counterpart. Third, while increasing the number of pipeline stages results in a descending trend in throughput per Watt for RAPID multipliers and accurate IPs (both multiplier and divider), this trend is ascending for RAPID dividers. This means that increasing the pipeline depth is highly beneficial for dividers. Moreover, the *relative* throughput per Watt ratio for RAPID dividers is higher than 1, meaning that pipelined RAPID dividers achieve a better throughput per Watt than their accurate counterparts (even when compared to the non-pipelined mode). On the other hand, as discussed in Chapter 3, the reciprocal-based dividers are not suited for LUT-based platforms. In fact, the poor performance of pipelined SAADI-EC is due to two reasons: First, its data path is divided into three non-uniform stages (normalizing block, multiplier, and error-correction accumulator based on Taylor iterations). Second, generating the reciprocal of the divisor (even through utilizing reciprocal IP of Vivado, as adopted herein) is a costly operation for LUT-based designs.

It is worth noting that Xilinx Vivado offers different implementation strategies to reduce the dynamic power, including Power_DefaultOpt, Power_ExploreArea (in which sequential resources are combined) and system-level power reduction techniques such as voltage scaling [Xil21]. Overall, it is reported that such optimizations can enable up to 30% improvement in dynamic power [Xil21]. However, considering the trade-off between area/power and performance, exploiting such techniques will increase the critical path delay. As pipelining has increased the critical path delay by itself, further directives were not applied, but such options can be utilized by users with respect to their constraints.

The Resource Usage for the Error-Reduction Schemes: From the hardware-usage perspective, each 6-LUT inherently functions as a 4-MUX (in which 4 inputs and the 2 select lines are the 6-LUT inputs). Further, as also denoted by multiplexer-implementation reports of Xilinx [Cha13], a 16:1 multiplexer requires a single FPGA-slice having four 6-LUTs. Based on this, a squarish partitioning was also chosen through a MUX-based approach to minimize the overhead of the proposed error-reduction strategy. On the contrary, checking the exact rhombus borders (see the original shape of the error in Fig. 2 of the paper) is very costly in hardware and could nullify the LUT-saving that is gained from the approximating approach.

The Scalability/Efficiency of the Proposed Scheme: The proposed partitioning is also scalable, as contrary to the REALM and SIMDive approaches (in which the number of coefficients and it's associated resource usage is increased exponentially), RAPID only uses few error-coefficients with small MUXes. In fact, the proposed partitioning surpasses the state of the arts in terms of resource-error trade-off: With 10 error coefficients and consideration of 4 MSBs, an ARE of 0.6% is achieved, which is better than SIMDive/REALM (0.8% ARE, considering 3 MSBs). Moreover, the resource footprint of this scheme is 193 LUTs, still lower than that of the SIMDive/REALM counterparts.

Clock Power and Tput/Watt of Pipelined Versus Non-pipelined Designs: The clock power is non-negligible, especially for low bit-width, but this overhead reduces when increasing the bit-width. And this is because, in lower bi-with precision, there is less logic relative to the clocked circuit. It should be also noticed that this observation also holds true for the pipelined accurate IPs, provided by Vivado. However, the clock power of RAPID multipliers is 21% (for 8-bit) to 51% (in 16-bit) less than their accurate pipelined counterparts. In fact, although the Tput/Watt is decreasing for the pipelined *multipliers* (both accurate and RAPID), the relative Tput/Watt in RAPID designs is higher than their pipelined accurate counterparts. As discussed before, clock gating has been applied to prevent superfluous switching activity in unused resources. Beside that, Xilinx Vivado offers different implementation strategies including Power_DefaultOpt and Power_ExploreArea. In these optional optimizations, the sequential resources of the design are combined, especially to reduce the dynamic power [Xil21]. However, considering the trade-off between power and delay, exploiting such directives will further increase the critical path delay. As pipelining already increases the critical path delay by itself, further directives were not applied, but such optimizations can be utilized by designers with respect to their constraints. In addition, there are also system-level power reduction techniques such as voltage scaling and clocking at a lower frequency. Overall, it is reported that such optimizations can enable up to 30% improvement in dynamic power [Xil21]. It should be noted that analyzing such techniques is beyond the scope of this paper, but opens up interesting topics which can be pursued in the future tracks.

4.6.4 Results for SIMDive Multipliers and Dividers

The following observations can be deducted based on the circuit-level results of SIMDive, summarized in Tables 4.7, 4.8, and 4.9.

Performance metrics of SIMDive (SISD mode) compared to baseline multipliers and dividers: As can be seen in the tables, SIMDive designs yield a better

Table 4.7 Accuracy-resource trade-off of SISD approximate multipliers in the FPGA implementation

Multiplier							
8 × 8	Area (6-LUT)	Latency (ns)	Power (mW)	Energy (uJ)	ARE (%)	PRE (%)	Error bias (%)
Acc-IP NP [EUK20]	72	5.2	15.10	79.06			
DRUM-4 [SIS15]	**53**	**4.57**	15.15	**69.37**	5.82	25.35	1.84
AFM1 [YHS20]	69	5.5	14.68	76.60	**0.23**	16.52	0.23
Mitchell [NMi62]	**51**	**4.81**	**13.21**	**66.03**	3.77	11.11	3.77
MBM [HHS18a]	64	5.16	13.36	71.54	2.60	8.59	0.09
LeAp-5 [1][ZSA20]	56	4.93	14.82	73.67	0.99	4.96	0.06
SIMDive	57	5.08	**13.72**	74.72	**0.82**	**4.76**	**0.05**
16 × 16	Area (6-LUT)	Latency (ns)	Power (mW)	Energy (uJ)	ARE (%)	PRE (%)	Error bias (%)
Acc-IP NP [EUK20]	287	7.04	45.20	311.90			
DRUM-6 [SIS15]	233	**5.34**	42.59	232.24	1.47	6.31	**0.04**
AFM1 [YHS20]	261	7.32	37.98	257.20	1.34	17.80	1.34
Mitchell [NMi62]	**187**	**4.9**	**34.15**	**177.2**	3.85	11.11	3.85
MBM [HHS18a]	204	6.59	35.92	214.76	2.63	8.83	0.09
LeAp-5 [ZSA20]	227	6.15	36.09	191.71	0.98	4.80	0.06
SIMDive	**216**	5.94	**34.82**	**184.24**	**0.82**	**4.90**	0.05
32 × 32	Area (6-LUT)	Latency (ns)	Power (mW)	Energy (uJ)	ARE (%)	PRE (%)	Error bias (%)
Acc-IP NP [EUK20]	1037	9.81	114.32	1065.4			
DRUM-6 [SIS15]	616	**6.35**	77.60	356.8	1.53	5.88	**0.05**
AFM1 [YHS20]	995	10.76	87.94	815.9	2.88	22.40	2.88
Mitchell [NMi62]	**484**	**5.96**	**57.1**	**334.3**	3.91	11.11	3.91
MBM [HHS18a]	533	7.51	60.14	447.6	2.69	8.74	0.10
LeAp-5 [ZSA20]	547	7.19	60.25	411.5	1.03	4.87	**0.05**
SIMDive	**521**	6.88	**58.32**	**362.0**	**0.91**	**4.72**	**0.05**

[a] Initial version of LeAp is a sequential design. For a fair comparison, it is implemented combinational-based herein

performance-accuracy trade-off than other counterparts. In fact, although the gains from SIMDive designs are highly pronounced in 32- and 16-bit, the performance metrics of both SIMDive multiplier and divider are still superior to accurate IPs in 8-bit precision. In particular, the delay and energy in the proposed divider are improved by $\sim 4\times$ and $\sim 4.6\times$, respectively, compared to the accurate counterpart. This means that the division's long latency is bridged nearly to the latency of a same-size multiplier. In contrast, the SAADI-EC [Mel+19] implementation approach dissipates even more energy with lower throughput than the accurate divider. The improvement in SIMDive designs is due to the transformation of the 2D array-based

Table 4.8 Accuracy-resource trade-off of SISD approximate dividers in the FPGA implementation

Divider								
8/4		Area (6-LUT)	Latency (ns)	Power (mW)	Energy (uJ)	ARE (%)	PRE (%)	Error bias (%)
Acc-IP [Xil16b]	NP	51	10.74	11.82	128.44			
AAXD [HJi+19]	6/3	38	6.06	12.21	75.65	2.08	100	1.49
AAXD [HJi+19]	4/2	**30**	**4.28**	**10.06**	**51.72**	3.74	100	2.58
Mitchell [NMi62]		**36**	**5.1**	**11.45**	59.3	3.9	13	3.9
INZeD [HHS19]		47	6.12	14.53	81.03	2.93	9.53	**0.02**
SAADI-EC [Mel+19]		103	12.8	17.50	214.50	2.37	8.82	1.92
SIMDive		44	5.33	12.24	67.60	**0.77**	**5.2**	**0.01**
16/8		Area (6-LUT)	Latency (ns)	Power (mW)	Energy (uJ)	ARE (%)	PRE (%)	Error bias (%)
Acc-IP [Xil16b]	NP	169	21.67	24.11	537.6			
AAXD [HJi+19]	12/6	207	18.87	**21.26**	443.94	**0.74**	100	0.3
AAXD [HJi+19]	8/4	151	12.51	**19.53**	286.09	2.99	100	0.9
Mitchell [NMi62]		**122**	**5.38**	21.7	**112.5**	4.11	13	4.11
INZeD [HHS19]		165	6.28	23.41	145.05	2.93	9.54	**0.02**
SAADI-EC [Mel+19]		362	25.7	29.24	890.29	2.14	8.82	1.76
SIMDive		**143**	**5.56**	23.03	**123.90**	**0.78**	**5.2**	**0.01**
32/16		Area (6-LUT)	Latency (ns)	Power (mW)	Energy (uJ)	ARE (%)	PRE (%)	Error bias (%)
Acc-IP [Xil16b]	NP	597	46.14	33.08	1479.9			
AAXD [HJi+19]	12/6	513	37.2	**30.36**	1048.05	**0.79**	100	0.35
AAXD [HJi+19]	8/4	361	24.66	**26.71**	688.79	3.04	100	1.1
Mitchell [NMi62]		**349**	**6.11**	33.9	**236.1**	4.19	13	4.19
INZeD [HHS19]		422	8.15	36.56	305.47	2.96	9.47	**0.03**
SAADI-EC [Mel+19]		822	51.6	42.61	2143.33	2.33	9.04	1.85
SIMDive		**381**	**6.34**	39.94	**246.66**	**0.81**	**5.16**	**0.02**

Table 4.9 Design parameters for SIMDive hybrid and SIMD modes

	Area (6-LUT)	Latency (ns)	Power (mW)	Energy (uJ)	ARE (%)	PRE (%)	Error bias (%)
8-bit Hybrid MUL/DIV (SISD)	87	5.57	25.69	144.9	0.82	5.2	0.05
SIMD MUL/DIV (16/8 bit)	283	6.13	79.2	502.41			
	Area (6-LUT)	Latency (ns)	Power (mW)	Energy (uJ)	ARE (%)	PRE (%)	Error bias (%)
16-bit hybrid MUL/DIV (SISD)	261	5.84	57.33	235.9	0.82	5.2	0.05
SIMD MUL/DIV (32/16/8 bit)	1059	7.92	97.81	778.1			

multiplication (division) structure to 1D addition (subtraction) through Mitchell's algorithms. Moreover, the transformation between hybrid functionalities also comes with a tolerable cost. In fact, our hybrid modules (both SISD and SIMD modes) are still smaller and faster than one accurate multiplier.

Performance Metrics for SIMDive (SIMD Mode) Compared to Baselines: It is worth noting that the improvement obtained by SIMDive designs are also reflected in SIMD modes: For example, transforming our proposed integrated multiplier-divider from 16-bit SISD to 32-bit SIMD has increased the area by a factor of ~ 4. This factor is also ~ 3.7 for hierarchical designs, which operate only at an SISD mode. The reason behind such a low overhead is two-fold: (1) As discussed before, the detection of leading one in SIMDive is performed in parallel in each 4-bit segment of inputs, which is inherently suited for parallelization in a SIMD design. This enables resource reuse and imposes a small overhead when 32-bit LOD is converted to smaller instances. (2) Mitchell's algorithm is also inherently more suited for SIMD architecture, as converting SISD to SIMD in addition steps of the algorithm poses small overhead (modifying 32-bit SISD adder in fractional part to four 8-bit adders in SIMD mode).

Accuracy of SIMDive approach compared to baselines: SIMDive designs render higher accuracy than other counterparts in terms of ARE and/or PRE. In fact, there are a lot of cases with errors near or equal to 100% in the dynamically truncation-based approach of the truncated-based AAXD divider (that also increases with truncating more bits). Such high error cases can result in false positive peaks in heartbeat and corner detection applications, which are discussed in Sect. 6.4. Moreover, the unbiased errors of SIMDive can cancel out each other and minimize error accumulation in the aggregation-based applications, having mainly addition/multiplication operations (also true for bio-signal and image-processing domain). Furthermore, recalling

Sect. 4.4, boosting the accuracy in the SIMDive approach comes with a negligible cost, compared to the original Mitchell's design: One more 6-LUT increases the bit-width of the error-coefficient by one bit (in all 64-coefficients).

4.7 Conclusion

This chapter presented various configurations for approximate multiplier and divider units aimed at improving throughput, area, and energy-efficiency in FPGAs while maintaining a higher accuracy-performance trade-off than the cutting-edge counterparts. Our designs are based on the Mitchell multiplication and division algorithms and offer different accuracy-performance trade-offs. For example, RAPID designs enable a fine-grained pipelining for multiplication and division, while SIMDive unit enables the hybrid functionality of multiplication and division on-the-fly, without the need for reconfiguration. Moreover, SIMDive enables the data-level parallelism approach through supporting the SIMD mode. Each of these designs is then suited for specific execution approaches. For example, the SIMD mode of SIMDive is particularly suitable for matrix-multiplication included applications. It is also worth highlighting that the lightweight error-reduction approaches of the proposed designs have been able to bound the errors to $< 1\%$. The error-bias is also minimized after applying the proposed error-reduction approaches. Such a feature plays an important role in minimizing the accumulated errors, in the approximation of multi-kernel applications. The efficacy of our proposed approximate multipliers in various applications was also evaluated, the results of which will be presented in Sect. 6.4.

Chapter 5
Architecture-Level Approximations

Abstract This chapter explores edge computing demands driven by diverse Internet of Things (IoT) applications, from health monitoring wearables to real-time video and image processing . Given that 47% of cardiac incidents occur outside hospitals [Wor18, Kos+17], adaptive low-power computation becomes crucial. The proliferation of UAVs and video-based services, which already account for over 82% of global internet traffic [Cis20b, San21], further emphasizes edge-side processing efficiency. CGRAs offer a promising trade-off between flexibility and energy efficiency via word-level reconfigurability, enabling customization to patient-specific conditions or evolving applications. This chapter presents two approximate CGRA architectures offering precision-adaptable, function-versatile PEs that enhance area, performance, and energy efficiency. Through multi-level approximation, including precision scaling and inexact arithmetic, and employing the SIMDive hybrid multiplier-divider , the proposed designs demonstrate scalable approximation control suitable for robust and energy-aware edge computation.

Keywords Cross-layer approximation · FPGA · CGRA · SIMD · MIMD · High-throughput · Edge computing · Sustainability · Energy-efficiency · Area optimization · Mitchell's algorithm · Logarithmic hybrid multiplier-divider

5.1 Introduction

The ever-growing demand for computing at the edge, for a wide domain of applications (from bio-signal to various cutting-edge image processing), has become more pronounced in the era of IoT. For example, wearable 24/7 health monitoring gadgets are becoming ubiquitous, especially noting that 47% of cardiac diseases—the main cause of death worldwide—occur outside of hospitals [Wor18, Kos+17]. Moreover, real-time video streaming and image processing are ubiquitous in various IoT services, UAVs such as drones are also proliferating for, e.g., object/self-tracking, search and surveillance, agricultural operations, and entertainment. In fact, Cisco has reported that the share of video processing traffic in the global internet traffic has

Z. Ebrahimi and A. Kumar, *Cross-Layer Approximation and In-Network Acceleration*,
https://doi.org/10.1007/978-3-032-21712-7_5

already passed 82% and is projected to increase in upcoming years [Cis20b, Cis18] and continues to grow in the upcoming years [San21, San19].

On one hand, the hardware platform for implementing such applications should be customizable w.r.t many conditions. For example, the health gadgets should be able to adapt to various patients' physiological traits[1] and the changes in heart's activity.[2] On the other hand, the resiliency of computation algorithms to error in these applications motivates the use of approximation techniques to improve energy efficiency, which is of utmost importance for such battery-powered devices. For example, approximation potentials are highly pronounced in bio-signal analysis, as (1) *processing* stage dissipates up to 70% of total energy in wearable nodes [Bra+17]. (2) These signals are inherently subjected to noise, and their processing algorithms exhibit high parallelism and error-resiliency [PRS19b]. Therefore, such computations can benefit greatly from an approximate CGRA that is adjustable with the user requirements, e.g., patient's changing condition/activity/physiology or application upgradability (which usually outpace hardware updates).

CGRAs can strike a good balance between flexibility and energy efficiency (recalling Fig. 1.2 in Sect. 1.5) through word-level reconfigurability, making them viable acceleration platforms for executing a broad range of tasks at the edge. In fact, CGRAs are already commercialized in Samsung Galaxy [Kim+14], Intel datacenter processors [Tan+19, Bar+15], IBM's RaPiD which is an approximate AIs accelerator [Ven+21b], Wave Dataflow Processing Unit (DPU) for Deep Nural Networks (DNNs) [Nic17] or persuaded the major FPGA vendors to coarsen the granularity of their chips [Gai+19], [FGS19].

This chapter presents two novel approximate CGRA architectures that target area, performance (throughput), and energy efficiency. This is achieved through designing light-weight PEs, each of which enables function-versatility and precision-adaptability by supporting a chain of approximations across the layers of abstraction (i.e., precision scaling on top of inexact multiplication and division). In this context, the approximation strategy of the SIMDive hybrid multiplier-divider was adopted, which can also perform sub-word parallelism through its SIMD execution mode. The novel contributions of this chapter are listed as follows.

- BioCare CGRA is presented, in which each PE implements a chain of approximations across layers of abstraction (i.e., precision scaling on top of inexact multiplication and division). By capitalizing on the SIMDive-based hybrid multiplier-divider (which is also capable of performing sub-word parallelism through its SIMD execution mode), BioCare supports function versatility and precision adaptability within a small area footprint.
- The function versatility and precision adaptability of SIMD PEs in BioCare (16/8-bit) are expanded to also support a MIMD mode in the proposed GREEN CGRA, for 16/8/4-bit precision.

[1] Bio-signals such as Electrocardiogram (ECG)/Electroencephalogram (EEG) vary person-to-person and also in each person, age-to-age.

[2] Many abnormality symptoms also exist 24h prior to a deterioration [Anz+17].

- The architecture of GREEN is equipped with a global memory, which enables measuring the reconfiguration time and energy for the end-to-end evaluation of multi-kernel applications on the GREEN CGRA.
- The BioCare and GREEN CGRAs are evaluated on eight single- and multi-kernel case studies from different domains: classification, bio-signal (ECG/EEG), and image/video processing.

5.2 BioCare: Energy-Efficient SIMD CGRA for Bio-signal PRocessing at the Edge

To design the BioCare CGRA—which enables an accuracyâŁ"energy trade-off—approximation and optimization techniques were targeted at three layers of abstraction: Inexact multiplication and division (circuit-level) [EUK20], utilizing SIMD opportunities (architecture-level), and precision scaling (application-level) [Ebr+21]. The reason behind adopting precision scaling is three-fold: First, not only does it reduce the computation load, but the application's end-to-end latency may be shortened due to the reduction in the propagation delay of individual operations. Second, precision scaling perfectly aligns with the structure of SIMD modes (when needed) and can further reduce the memory footprint and its associated data movement energy. Third, as shown by multiple studies on the same applications, neglecting the LSBs due to the precision-scaling will not significantly affect the output accuracy of the application [PRS19b, HHS19, Ebr+21].

We adopted the approximation of multiplication and division operation for the circuit-level techniques via the SIMDive approach discussed in Sect. 4.4. The rationale behind using this approach is multi-fold: ① SIMDive error-reduction strategy is adjustable (by changing the number of error-reduction coefficients) and independent from the operand size. Moreover, SIMDive errors are nearly unbiased, meaning that they are centered and nearly symmetrical around zero. Therefore, the errors can cancel out each other in consecutive kernels with an aggregation-based structure (i.e., having mostly addition/multiplication operations) and prevent accuracy degradation [Zer+21, SAn+20, Tas+20]. Such a low-error feature also creates the opportunity to apply further approximation techniques, e.g., precision scaling on top of approximate multiplication and division. ② SIMDive supports various precision of division. This enables offloading division-included kernels to the accelerator and amortizes communications with the host processor. Moreover, performing SIMD division has become possible via the approximation approach of SIMDive, which is not feasible for the *accurate* mode. Furthermore, the SIMDive division approach bodes well for the design of ALU, as division acts as the speed-bottleneck operation in ALU. Therefore, integrating the multiplier and divider units not only substantially circumvents the long latency of division to the latency of the same-size multiplier but also significantly reduces the critical path delay of ALU. ③ SIMDive has shown superior resource-accuracy trade-off over state of the arts [ZSA20], due to converting the 2D structure of array-based multiplier (divider) to 1D shift and add (sub) in the logarithmic representation. In addition, SIMDive also enables on-the-fly switching between two functionalities (without the need for reconfiguration). This further facilitates a resource-efficient implementation for an ALU, preventing the overhead of collecting individual units separately. Moreover, the logarithmic conversion approach of SIMDive is also suitable for designing an SIMD architecture: Supporting a higher-precision mode is achieved by connecting smaller instances with a modest overhead. In fact, enabling the sub-word-level parallelism for the SISD version of multiplication/division operation (to also provide support for SIMD mode) is facilitated by configuring the carry chain in the adder/subtractor unit, LOD units, multiplexers associated with small Read Only Memorys (ROMs), and output barrel-shifters. We have also designed the ASIC-based implementation of the SIMDive (which originally was proposed and implemented with an LUT-customized approach [EUK20]) to embed it in our SIMD CGRA. Another modification of ASIC-based SIMDive over the FPGA-based alternative also lies in reducing the error coefficients for multiplication mode (from 64 to 32), as discussed in the following.

5.2.1 Structure of BioCare SISD/SIMD CGRA

Figure 5.1 shows the architecture of BioCare CGRA [EK21]. Table 5.1 also details the proposed ALU opcodes in BioCare CGRA that support different levels of precision on the fly. The presented eight opcodes are selected, based on the most frequent operations that are required in the case study applications. This selection has been done considering both the results of the sensitivity analysis (that reveals the lowest bit-width precision for each of the kernels' operations, at which the accuracy is still acceptable) and profiling the sequence of operations in the computational kernels. Nonetheless, opcodes can be customized or expanded depending on the requirements of the target application, especially considering that 10-bit immediate field (rather than 16-bit in the current version) suffices accommodation of the largest filter-coefficients in the case study applications. The BioCare ALU was operating at the precision of 16-[3] and 8-bit, as it was originally designed for processing of bio-signals, which are usually sampled at 8 to 16-bit and satisfy the accuracy-energy requirements of battery-powered remote monitoring gadgets (especially as many abnormality symptoms exist 24h prior to deterioration [Anz+17]).

Structure of PEs Each PE in the 2D array structure of BioCare also encompasses an exclusive context memory that configures the ALU for SISD/SIMD modes. The ALU is assembled by a shifter, two LOD units (to determine the position of the leading one in the binary representation of the input), an adder, and a two's complement unit which handles the signed numbers or subtraction operation (also used in SIMDive division). Performing the *log* is obtained by detecting the leading-one and then extracting the fractional-part (using a barrel-shifter of appropriate sizes, as shown in

[3] Note that connecting the output of multiplication results requires 32-bit.

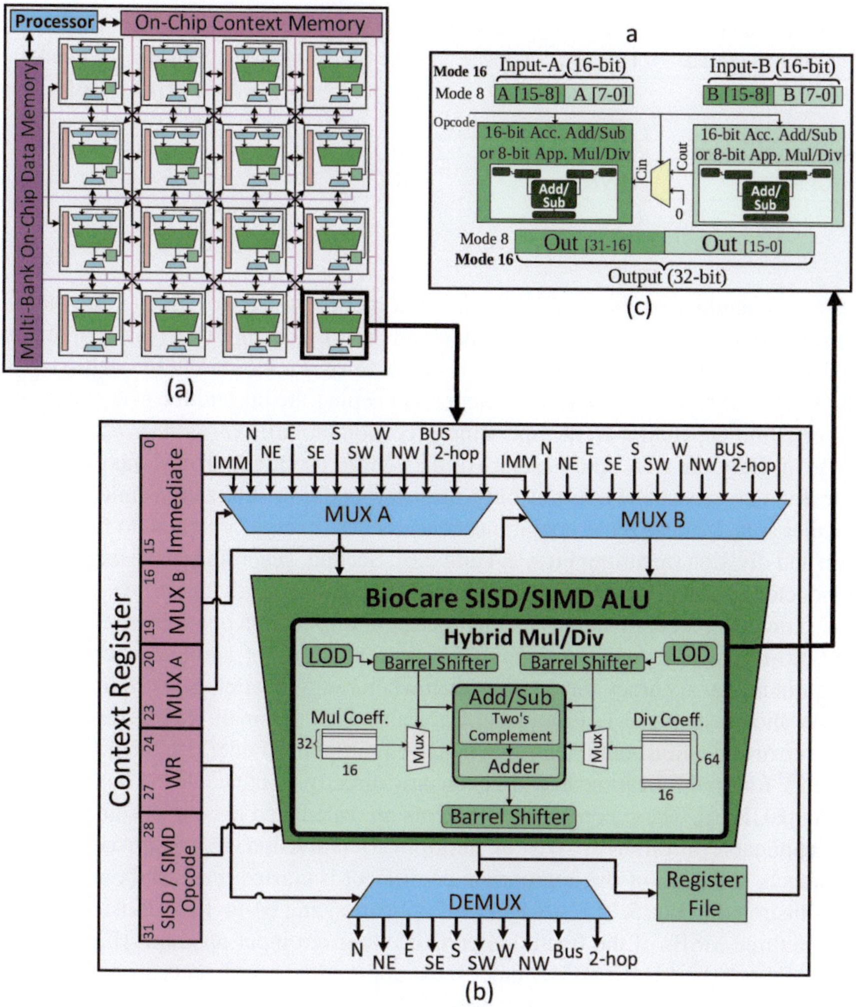

Fig. 5.1 **a** BioCare CGRA, **b**, **c** Proposed SISD/SIMD PE/ALU

Fig. 4-b of the paper). Similarly, for the *anti-log*, another barrel-shifter is used which scales the summation (subtraction) of error-correction term and fractional parts w.r.t. the summation (subtraction) of integer parts, depending on the multiplication (division) mode. For example, when operation precision is 16-bit, the maximum bits to be shifted for the input and output barrel-shifters are 15 and 31, respectively, in multiplication mode, as the summation of error-correction terms sometimes results in an overflow: $integer\_part1 + integer\_part2 + overflow = 15 + 15 + 1 = 31$. The barrel shifters use the behavioral description approach (case statements and shift operators). During the shifting, extra zeros are inserted to the right side of the fractional parts for proper alignment and keeping the bit-width constant (i.e., 4-, 8-, or

Table 5.1 Supported SISD and SIMD opcodes in BioCare CGRA

Function	Opcode	Function	Opcode	Function	Opcode	Function	Opcode
ADD32	0000	MUL16	0001	DIV16	0010	ADD16_ ADD16	0011
MUL8_ MUL8	0100	MUL8_ DIV8	0101	ADD16_ MUL8	0110	ADD16_ DIV8	0111

16-bit, depending on the operating-precision mode). In term of coding, the barrel-shifters are implemented using the behavioral description approach (case statements and shift operators). During the shifting, extra zeros are inserted to the right-side of the fractional parts for proper alignment and keeping the bit-width constant (i.e., 4-, 8-, or 16-bit depending on the operating-precision mode).

In the BioCare ALU, the structure of the addition/subtraction unit was kept accurate for two reasons. First, not only does the addition-subtraction unit require a smaller area, but also these operations consume less energy compared to multiplication and division (recalling Figs. 2.4 and 2.6). Second, according to the experiments conducted in recent studies [HJi+20, Ebr+21] and our sensitivity analysis (see Sect. 6.4.3) on different applications having matrix multiplication operations (e.g., image and bio-signal processing domains), the approximation of addition operation can result in higher accuracy fluctuations than when multiplication is approximated.

As shown in Fig. 5.1, PEs in BioCare also include small ROMs for the constant error-reduction coefficients of SIMDive multiplier-divider [EUK20]. The coefficients for multiplication and division are directly obtained from our previous study [EUK20]. These constant coefficients are calculated once at design time and are applicable to different sizes of multiplication/division. In this error-reduction approach, selection of the appropriate coefficient is carried through the multiplexers illustrated in Fig. 5.1 (32-to-1 for multiplication and 64-to-1 for division), based on the three MSBs of the fractional parts of the given input operands [EUK20]. In case of 4 or 8-bit SIMD/MIMD modes, the appropriate coefficients are selected via separate multiplexers and added to each of the 4/8-bit ALUs.

A delicate optimization has also been adopted in the BioCare CGRA over the SIMDive [EUK20]: The coefficients of multiplication can be halved, as swapping operands does not affect Mitchell's error for this operation. Such a reduction has also reduced the complexity of the associated multiplexer that selects the error coefficient. Nevertheless, the ROM and its associated multiplexer do not impose a significant overhead in CGRA, especially as the multiplexer's complexity increases logarithmically. As CGRAs have variations of interconnect structures depending on the application domain, BioCare was augmented with diagonal and 2-hop links (for the third and sixth rows), enabling energy-efficient processing of bio-signals [LSF17]. Although these domain-specific requirements have increased interconnection area/power, ultimately, Instruction Per Cycle (IPC) and execution time/energy of the total application has improved: Prolonged schedules are avoided when PEs are not used as routing nodes. Similar observations are also drawn in [Bou+07, Yoo+13, Chi+17].

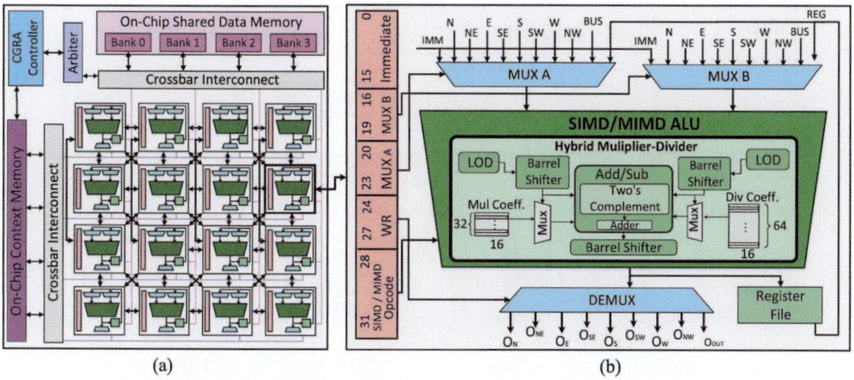

Fig. 5.2 **a** GREEN CGRA architecture, **b** The proposed structure of SISD/SIMD/MIMD PE

5.3 GREEN: Approximate SIMD/MIMD CGRA for Energy-Efficient ProcessiNg at the Edge

In this section, the architecture of GREEN CGRA is elaborated, which mainly expands the function versatility and precision adaptability of the BioCare CGRA. In fact, GREEN enables a cross-layer approximation approach through the flexibility of its PEs; each can concurrently perform addition, multiplication, *and* division on different data parts (4, 8-, or 16-bit data precision). This conjoins instruction- and data-level parallelism (ILP/DLP) into a light-weight realization of MIMD [Kim+09]. It should be noted that although all CGRAs belong to MIMD execution model (according to Flynn's taxonomy [Liu+19b]), GREEN has even taken a step further and enabled multi-functionality with varied precision within *each* PE. In fact, while SIMD mode can perfectly fit for the acceleration of similar operations with varied precision, e.g., convolutions, MIMD allows better utilization of PEs when simultaneous computations of varied operations are of interest [Kim+09]. Prevalent examples that can benefit from this light-weight SIMD/MIMD architecture are a broad range of matrix-multiplication inclusive applications, e.g., image, video, and bio-signal processing applications along with NNs . Interestingly, all these cases also involve division operations, which may require precision different from multiplication. The need for such SIMD/MIMD architectures has also been cited in recent industry-academia collaborations for emerging frontiers of AI acceleration [Yaz+18].

The Structure of GREEN SISD/SIMD/MIMD CGRA

GREEN CGRA, depicted in Figs. 5.2 and 5.3, features a systolic array structure containing homogeneous PEs, each of which encompasses an ALU that can be configured to SISD, SIMD, or MIMD mode.

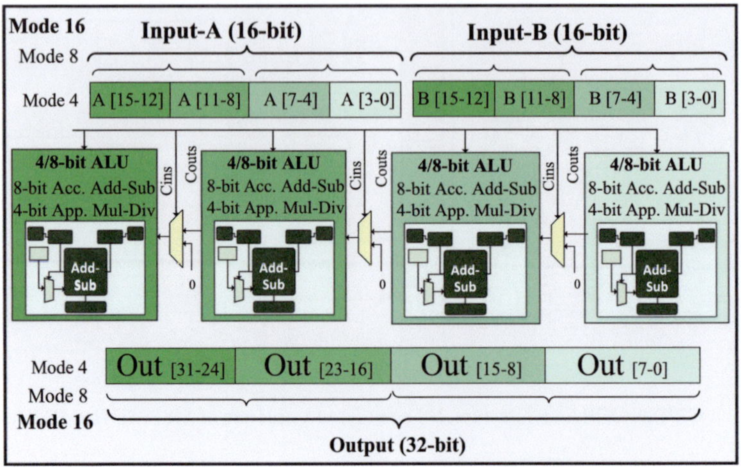

Fig. 5.3 The SISD/SIMD/MIMD structure of GREEN ALU (based on ASIC-modified multiplier-divider of SIMDive)

Structure of PEs Like BioCare, each ALU is mainly assembled by a shifter, two LOD units, an adder, and a two's complement unit to handle signed numbers or subtraction operations. Like BioCare, the addition/subtraction unit structure was kept accurate. Enabling sub-word-level parallelism via modifying the SISD, to also provide support for SIMD/MIMD modes, is achieved by configuring the carry chain (for Add/Sub units), LOD units, multiplexers associated with small ROMs, and output barrel-shifters to provide support from 16-bit SISD for parallel 4/8-bit modes. For the context memory, the size was considered such that it can accommodate all required configurations for the eight single- and multi-kernel case study applications introduced in Sect. 2.6. In fact, the required amount is much less than $2^{32} = 4GB$. The configurations of the 32-bit context words is as follows: {Opcode, WR, Mux-A, Mux-B, immediate}. The number of supported configurations are calculated depending on the supported modes by each field which can be calculated as follows: $\{16 \times 9 \times 11 \times 11 \times 2^{10}\} = 17842176$ bit, ~2.3 MB (10-bit immediate field suffices accommodation of the largest filter-coefficients in the case study applications).

Table 5.2 presents the implemented ALU opcodes in GREEN that supports on-the-fly alteration of operations' precision and functionalities in SISD/SIMD/MIMD modes (light, medium, and dark green, respectively). The presented 16 opcodes are selected based on the most frequent SISD/SIMD/MIMD functionalities that are required in the eight case studies from different domains of application (classification, image, and bio-signal processing). In fact, the selection has been done considering both the results of the error-resiliency sensitivity analysis (to obtain the lowest bit-width precision for each of the kernels' operations) and profiling the sequence of operations in the computational kernels of case study applications. Nonetheless, these opcodes can be customized/expanded depending on the requirements of the target application, especially considering that the 10-bit immediate field (rather than 16-bit in the current version) suffices for the accommodation of the largest filter coefficients in the case study applications. Like BioCare, PEs also include small ROMs for the constant error-reduction coefficients for multiplication and division. Selection of the appropriate coefficient is carried through the multiplexers illustrated in Fig. 5.2 (32-to-1 for multiplication and 64-to-1 for division), based on the three MSBs of the fractional parts of the given input operands [EUK20]. In the case of 4 or 8-bit SIMD/MIMD modes (see Fig. 5.3), the appropriate coefficients are selected via separate multiplexers and added to each of the 4/8-bit ALUs. Finally, the intermediate result of a PE is stored in its local registers and can be accessed for the next operation.

Modifications to Enable Sub-word Parallelism at Run-Time Enabling the 16-bit SISD ALU to also provide support for 8-/4-bit SIMD/MIMD modes (shown in Fig. 5.3) is achieved by modifying each sub-modules, separately: The LODs and barrel-shifters are implemented with a modular approach based on 4-bit LODs. Together with extra circuitry (including a multiplexer which is controlled by the SISD/SIMD/MIMD signal), they can operate at 8- and 16-bit at run-time as well. Extra cascading logic is also used to concatenate the fractional parts when precision is 8- or 16-bit. On the other hand, the adders became configurable to 4-, 8-, and 16-bit by inserting multiplexers at proper locations along the carry chain (see Fig. 5.3). As mentioned previously, the error-coefficients are applicable to different sizes of multiplication/division, therefore, marginal modification were required for the ROMs, so that they can be accessed by multiple ports (in case of sub-word parallelism). Run-time adjustment to different approximation modes (i.e., precision-scaling or word-level parallelism) can be achieved in two steps. First, for each of the PEs the proper context word is from the context memory, in which the SISD/SIMD/MIMD modes of the PEs is determined. Second, the ALUs inside PEs are configured w.r.t. the PEs context word. For example, the carry chain in the SIMDive units can be configured to support word-level parallelism.

Local and Global Memories In the GREEN architecture, two memory types are used. The local memories are dual-ported register files, and the global memory is constructed out of dual-ported Static Random, Memory Access (SRAM) units. The size of global memory is 12 KByte and it is sufficient for, e.g., a 30-second batch of 16-bit ECG samples acquired at a sampling frequency 200 Hz (30 sec $\times$ 200 Hz $\times$ $\frac{16}{8}$ Byte = 12 KiB). Each bank in the global data memory can be individually powered off, on, or placed into retention mode until the next processing interval. The multi-banking feature also allows performing simultaneous operations with various functionalities and precision needed for SIMD/MIMD modes. Currently, the GREEN CGRA loads data from global memory and stores it in the local register files, and the design of a DMA Controller is envisioned for future versions of GREEN.

Network Structure The interconnect pattern in GREEN CGRA is a 2D mesh. As CGRAs have variations of interconnect structures depending on the application-domain use case, the network was augmented with diagonal and 2-hop links (between the first and third rows), enabling energy-efficient processing of bio-signals [LSF17]. Although adding these links has increased the interconnection area/power, ultimately, the IPC and execution time/energy of the total application has improved: Prolonged

Table 5.2 Supported SISD, SIMD, and MIMD opcodes in GREEN CGRA

Function	Opcode	Function	Opcode
ADD32	0000	DIV8_DIV8	1000
MUL16	0001	DIV4_DIV4_DIV4_DIV4	1001
DIV16	0010	ADD8_MUL4_MUL4_MUL4	1010
ADD16_ADD16	0011	ADD8_ADD8_DIV8	1011
ADD16_ADD8_ADD8	0100	ADD8_MUL8_DIV4	1100
ADD8_ADD8_ADD8_ADD8	0101	ADD8_MUL4_DIV8	1101
MUL8_MUL8	0110	MUL8_DIV4_DIV4	1110
MUL4_MUL4_MUL4_MUL4	0111	ADD16_MUL8	1111

schedules are avoided when PEs are not used as routing nodes. Similar observations on the benefit of using diagonal links are also drawn in [Bou+07, Yoo+13, Chi+17].

Controlling the CGRAs In our current implementations, there is no dedicated or extended ISA for controlling the array, that typically would be issued by the processor in an system on Chip (SoC) architecture. The controlling process (detailed below), in our setup, is managed by the CGRA controller though calling various functions—with a proper order—in the written C-codes: (1) scheduling and binding (mapping) (2) data initialization in the local memories (3) configuring the PEs, i.e., functionality and precision as well as the connections between PEs, through the context words (4) writing back to the global memories and reconfiguration. These tasks are discussed as follows:

Scheduling and binding (mapping): Similar to the X-CGRA work [Akb+20], the mapping phase was separated into two steps: scheduling and binding. At first, the start time of (group of) operations in the DFG are determined, via employing the list scheduling algorithm, in a way to preserve a correct execution order considering the operations dependency. Afterwards, for the binding, the DFG nodes are sorted according to their time stamp and the graph-coloring mapping (where no two adjacent DFG nodes have the same color) is performed in a node-by-node fashion in a while loop: For each node and unmapped PE, the path with minimal cost is selected using Dijkstra's shortest path algorithm [Dij22], during which some PEs are inevitably used as routing nodes when no direct connection exists to implement two PEs with data dependency.

Configuring PEs: The functionality and precision of operations (ALU opcode and SISD/SIMD/MIMD mode) for all PEs, is determined through assigning the context words by the controller, based on the results of the mapping. The 'WR', 'Mux-A', and 'Mux-B' fields of the context words also establish the connection between adjacent PEs. Once the context words are determined, they are loaded into the context memories of individual PEs through *calling a void write_function*.

Data movement and reconfiguration: The controller is also responsible for data transfer between the local and global memories. Due to the lack of direct access into the memory structure of the commercial model used in this work (similar to the Blocks work [Wij+19, WCK21]), we only measured the energy and time for memory accesses in our evaluations. Design of a DMA unit for automating the data transfer (and relieving the host processor) is envisioned as part of the future work.

5.4 Architecture-Level Results and Discussion

5.4.1 Experimental Setup

We have assessed the collective performance metrics of BioCare and GREEN against state of the art CGRA for different sizes of PE array. BioCare is evaluated in both SISD and SIMD modes with 16-bit input (subword of 8-bit), while GREEN is assessed in all SISD/SIMD/MIMD modes with 16-bit (subword of 8/4-bit). state of the art approximate X-CGRA [Akb+20], GP-CGRA [Bra+18], are also implemented with the input bit-width similar to BioCare and GREEN. As there are multiple choices for inexact adder and multiplier units in the baseline studies, we analyzed the ones with the lowest *resource×error bias*. In addition, we have implemented the XBioSip approach [PRS19b] (it proposed an ASIC-based approximation architecture and methodology for bio-signal processing) by implementing it in a CGRA with a minimal overhead to make 2–16 LSBs of addition and multiplication configurable. For a fair comparison, we kept the same routing structure in all CGRAs and refrained from utilizing specialized scheduling/mapping optimizations. The divider size has been adjusted for 8-bit for all baseline CGRAs. This decision is adopted based on the reason that division is not only the most area- and energy-consuming ALU operation (recalling Fig. 2.6), but also it is the most sensitive operation w.r.t the changes in applications' accuracy (will be discussed in Sect. 6.4.3). In fact, our analysis delineates that some division-included kernels/applications undergo noticeable QoR drop when the precision of division is set to 4-bit. After setting the precision of division to 8-bit, we have analyzed approximate adder and multiplier candidates in each X-CGRA and GP-CGRA works and selected those structures with lowest *resource×error bias* (note, in X-CGRA and GP-CGRA studies, inexact adders and multipliers with different approximation approaches have been examined). It should be mentioned that the approximations of selected adders and multipliers are also set to the level that the 100% QRS detection is maintained for heartbeat detection applications. This is achieved by focusing more on approximating the multiplier units, as they are less QoR-sensitive and more energy-hungry than adders.

Architectures are coded from scratch in HDL Verilog, synthesized with Nangate 45 nm using Synopsis Design Compiler, placed, and routed with Cadence Innovus. All the memories are also implemented by commercial 40-nm technology that can be taped out [Wij+19, WCK21]. Register Transfer Level (RTL) simulations are performed on the post-synthesis net-lists to obtain the activity files and afterwards, the net-lists combined with the activity files are given to the Cadence Innovus for placement/routing. Critical path delay and power estimations are then obtained by simulating the post-placed-and-routed net-list at the typical corners. In this thsis, we have performed the synthesis of all CGRAs with the area-opt goal pragmas[4]: Compile ultra with -area_effort, -map_effort, or -auto_ungroup area option, set_max_area, In parallel, the accuracy of approximation approaches has been assessed in both circuit-level (ARE, PRE, and EB and application-level, i.e., PSNR, SSIM, RMSE, QRS detection percentage, number of correct vectors, and standard deviation error, through Python and MATLAB simulations.

5.4.2 PE-Level Results (Individual Adders , Multipliers, and Dividers)

Table 5.3 summarizes the circuit-level characteristics of inexact adders, multipliers, and dividers and provides insights for the selection of SIMDive [EUK20] which bodes well for design of an ALU to be utilized in the GREEN CGRA. The following inferences are highlighted in the results:

- Table 5.3 shows that the SIMDive multiplier/divider achieves the lowest EB while yielding higher performance improvement. Specifically, the latency of the divider is reduced to less than that of an accurate multiplier of the same size. This justifies that SIMDive hybrid multiplier/divider suits an approximate ALU design. It is also worth underlining that although the application-level analysis exhibits that even the 4-coefficient version of SIMDive provides 100% QRS detection accuracy, the more accurate 64-coefficient version was used to enable the adaptable greedy precision-scaling strategy of [Ebr+21] for multi-kernel applications.
- The circuit-level results also demonstrate that the Mitchell-based designs are more suited than accurate or modular-based competitors for an SIMD ALU design as: (1) an SIMD divider is only achievable in an *approximate* mode. (2) Considering the hierarchical-based structure of accurate multipliers (2D array of HA/FA), the resource footprint grows quadratic (x^2) when the operand size is doubled. This factor would be smaller (~2.8) in the SIMDive-based ALU: A double-sized adder/subtractor is achieved by connecting two smaller instances. Moreover, thanks to the 2D to 1D conversion through Mitchell's logarithmic multiplication , the adder/subtractor unit is re-used in the SIMDive multiplication-division unit.

5.4.3 Architecture-Level Results (the Whole CGRA Architecture)

Tables 5.4 and 5.5 detail the architecture-level metrics of different CGRAs at post-synthesis phase, which includes the share of *logic* and *logic plus routing*, respectively.

[4] Such pragmas also inevitably contribute to the power improvement of the circuit.

Table 5.3 Circuit-level metrics of 16-bit adder and multiplier and 8-bit divider (normalized to the accurate versions)

		ARE (%)	PRE (%)	EB (%)	Area	Power	Delay	Energy
Accurate	Acc_Add (16-bit)	-	-	-	1	1	1	1
	Acc_Mul (16-bit)	-	-	-	1	1	1	1
	Acc_Div (8-bit)[4]	-	-	-	0.6	1.13	1.28	1.44
GREEN BioCare	Acc_Add (16-bit)	-	-	-	1	1	1	1
	SIMDive Mul/Div 64-coeff. (16-bit) [EUK20]	0.8/0.7	6.9/5.2	-0.04/-0.01	0.51	0.48	0.71	0.35
XBioSip [PRS19b]	AppAdd5 (2–16)	0.01–30	100	-0.01-6.9	0.52	0.48	0.7	0.35
	AppMultV1[5] (2–16)	0.01–61	100	0.03–60	0.54	0.53	0.63	0.37
X-CGRA [Akb+20]	RAP-CLA W6	0.2	100	0.2	1.16	1.18	0.87	1.03
	Dadda DQ42C4	8.1	51	8.1	0.66	0.62	0.69	0.43
GP-CGRA [Bra+18]	Add_Design 4	0.3	100	0.01	0.57	0.45	0.66	0.35
	Mul_Version Lit	3.4	22	3.4	0.8	0.9	0.88	0.79

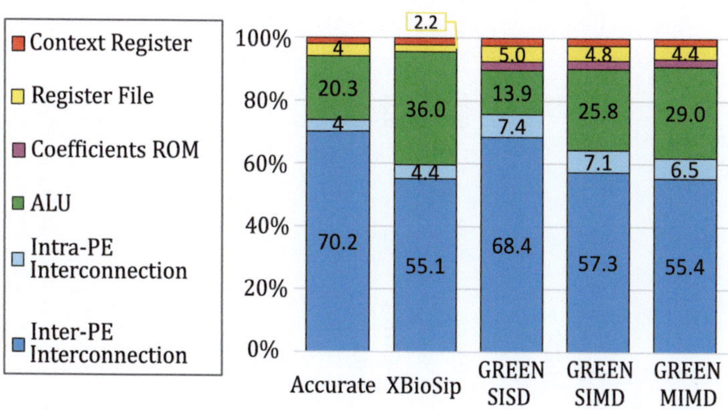

Fig. 5.4 Area breakdown of CGRAs after place-and-route (excluding the global memory). BioCare is the SIMD configuration of the GREEN CGRA.

Table 5.4 Architecture-level metrics of GREEN and baseline CGRAs at post-synthesis phase: GREEN MIMD mode represents the BioCare configuration

		Area[1] ($\mu m^2 \times 10^3$)	Power[2] (mW)	Chain Latency (ns)	Peak Through-put (GOPS)	Energy (pJ)	Area ($\mu m^2 \times 10^3$)	Power (mW)	Chain Latency (ns)	Peak Through-put (GOPS)	Energy (pJ)
		PE					**2×2**				
	Accurate	4.7	0.23	9.8	0.1	2.35	17.1	0.83	19.3	0.39	8.4
GREEN	**SISD**	3.4	0.2	4.9	0.2	1.16	11.5	0.72	10.8	0.74	4.4
	SIMD[4]	4.8	0.25	5.7	0.18–0.72[3]	1.43	17.4	0.9	11.3	0.7–2.8	5.7
	MIMD	5.2	0.27	5.9	0.17–0.68	1.6	18.9	0.97	12.3	0.65–2.6	6.7
	XBioSip	6.7	0.31	15.8	0.07	4.91	25.4	1.12	32.8	0.24	18.7
	X-CGRA	4.2	0.21	7.9	0.12	1.83	14.7	0.77	17.8	0.47	6.9
	GP-CGRA	4.6	0.22	8.5	0.11	2.07	16.6	0.8	18.3	0.44	7.6
		4×4					**8×8**				
	Accurate	71	3.5	39.4	1.59	33.7	279	14.2	79.2	6.27	141
GREEN	**SISD**	49	3.1	21.9	2.94	16.5	206	12.2	44.8	11.4	68
	SIMD	73	3.7	23.6	2.7–10.8	22.3	302	15.3	47.2	10.8–43.3	91
	MIMD	78	4	25.7	2.4–9.9	27	320	16.4	55	9.3–37	113
	XBioSip	104	4.6	67.4	0.95	78.9	423	18.9	154	3.5	364
	X-CGRA	61	3.2	36.1	1.72	28.7	245	13	72.5	7.2	118
	GP-CGRA	70	3.4	37.7	1.67	30.2	267	13.8	75.8	6.9	129

[1]Reports are for the logic, excluding the share for global memories and routing
[2]Synthesis of all CGRAs are performed with area-opt goal at 300 MHz frequency
[3]Peak throughput of SIMD/MIMD can be up to 4× (when operands are 4-bit)
[4]BioCare is the SIMD configuration of the GREEN CGRA

Table 5.5 Performance metrics of various CGRA architectures (including the shares of both logic and routing)

		Area (μm^2 $\times 10^3$)	**Power (mW)**	**Chain Latency (ns)**	**Peak Through-put (MOPS)**	**Energy (pJ)**	**Area (μm^2 $\times 10^3$))**	**Power (mW)**	**Chain Latency (ns)**	**Peak Through-put (MOPS)**	**Energy (pJ)**
		PE					**2×2**				
Accurate		13.6	0.73	28.9	3.46	77	55.9	2.79	91.9	13.4	299
GREEN	**SISD**	11.3	0.68	14.5	6.92	33.6	46.7	2.95	43.8	27.3	127
	SIMD	13.7	0.77	16.8	5.95–23.8	45.9	56	3.07	51.5	23.2–92.8	174
	MIMD	14.9	0.84	17.4	5.75–23	52.4	60.2	3.3	53.8	22.6–90.5	201
XBioSip		16.3	0.88	46.6	2.15	150.4	67	3.46	147.3	8.3	591
X-CGRA		13.1	0.7	23.3	4.29	57.1	54.3	2.82	72	16.8	223
GP-CGRA		13.5	0.71	25.1	3.99	63.7	55.1	2.84	77.7	15.7	249
		4×4					**8×8**				
Accurate		221	11	205	52.6	1147	922	47.4	465	207	4483
GREEN	**SISD**	186	11.6	102	107	492	777	45.5	224	419	1883
	SIMD	223	12	120	91.6–366	682	933	50.3	264	356 –1424	2627
	MIMD	241	13.2	124	88.3–353	793	1009	55.3	279	342–11368	3068
XBioSip		270	14.2	335	31.8	2353	1129	60	753	121	9271
X-CGRA		218	11.8	166	65.5	855	913	49.9	374	249	3324
GP-CGRA		219	12	180	61.8	952	916	53	402	237	3731

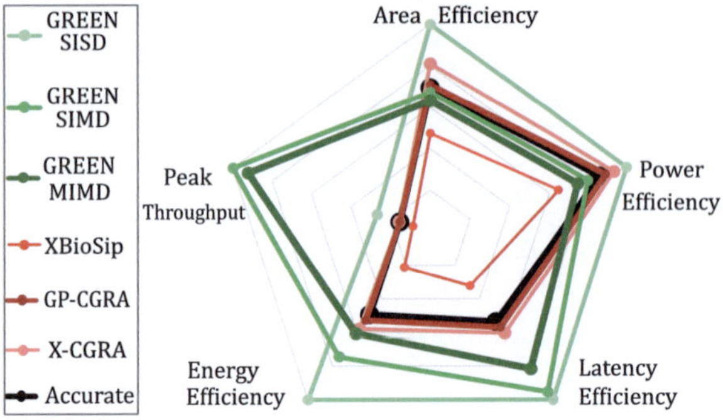

Fig. 5.5 Performance efficiency of CGRAs at architecture-level: GREEN MIMD mode represents the BioCare configuration (for relative comparison, efficiency is translated as the inverse of resource consumption). BioCare is the SIMD configuration of the GREEN CGRA

We have reported the share of logic separately, not only because the routing structure is the same in all CGRA, but also to highlight how the gains from various approximations at circuit-level (adder/multiplier/divider) changes, when they are embodied in the higher architecture-level. Nevertheless, our proposed architecture can ultimately reduce the used resources for routing as well. The reason behind is two-fold: (1) precision-reduced kernels implemented via the SIMD/MIMD, enable concurrent execution of more functions in the same PE and therefore, less number of PEs and therefore, less routing resources are required to implement a kernel. (2) featuring SIMD/MIMD at intra- rather than conventional inter-PE granularity mitigates stress on global routing.

To obtain the architecture-level results, the RTL simulations are performed on the post-synthesis net-lists for generating the activity files and afterwards, the net-lists along with the activity files provided the Cadence Innovus for placement and routing. Critical path delay and power estimations are then obtained by simulating the successfully placed-and-routed net-list at the typical corner at 1.0 V, 25 °C.
Please note, recalling Table 5.2, the SISD, SIMD, and MIMD versions of GREEN CGRA support 3, 10, and 16 customized ALU opcodes, respectively, and each variant renders different performance metrics. Afterward, Fig. 5.4 demonstrates the area-breakdown and performance-efficiency of all CGRAs after the placement and routing phase on the Cadence Innovus (in this figure, the SIMD version of GREEN is the BioCare CGRA). Finally, Fig. 5.5 illustrates the quantitative relation among the values (that are presented in Table 5.4) and compares the performance-efficiency of different CGRAs at the architecture-level. The following inferences are highlighted in the architecture-level results[5]:

- *GREEN outperforms other CGRAs at architecture-level*: As indicated in Table 5.4 and Fig. 5.5, the SISD version of GREEN is up to 32% smaller, 55% more energy-efficient, and 49% faster than the accurate CGRA. Although augmenting SISD to SIMD (i.e., BioCare) and MIMD structures has increased the complexity and therefore, resource consumption (e.g., 46.4% and 55.3% area-overhead for SIMD,

[5] Considering that the focus of this book is on the *approximation of CGRA*, the results are presented excluding the external memories.

Table 5.6 Maximum frequency for different CGRA architectures (MHz)

Accurate	SISD	SIMD	MIMD	XBioSip	X-CGRA	GP-CGRA
293	580	501	472	210	395	352

and MIMD modes, respectively, in the 8×8 CGRA), the maximum throughput is also significantly increased through SIMD and MIMD modes (up to 3.8× and 3.25× for SIMD and MIMD, respectively). Furthermore, the GREEN CGRA can enable up to 6.7× higher throughput compared to the accurate counterpart (when the PEs are configured to SIMD or MIMD mode and operate at 4-bit precision).

- Table 5.4 indicates that the XBioSip approximation approach [PRS19b] for multi-kernel applications (enabling accuracy-configurability of LSBs rather than precision-scaling of kernels) is beneficial, only in a fully-customized ASIC implementation. In contrast, the overhead for supporting such LSB configurability in the PE-based template of a CGRA is significant, all FAs, and 2×2 multipliers operating on LSBs of up to 16-bit should support accurate and approximate modes. As such, this approach will be counter-productive when realized in a homogeneous CGRA structure (please note, it is out of the scope of this work to go into designing and evaluating heterogeneous CGRA structures).
- Although the results presented in Table 5.4 and Fig. 5.5 demonstrate that GREEN SIMD has slightly better performance than the MIMD version at *architecture-level* (as having less complex PE due to supporting less opcodes, see Table 5.2), it will be shown that the MIMD mode renders better performance at *application-level* (see Fig. 6.22 in Sect. 6.4.3). The reason behind this is that the MIMD allows concurrent execution of more functions in the same PE, and therefore, the operations in the kernel require less number of PEs to be implemented.
- Although the maximum clock frequency for CGRAs ranges from ∼210 MHz to ∼580 MHz, following prior studies in the literature, the frequency was set to 300 MHz for all CGRAs (including synthesis and power analysis).

Table 5.6 below, details the maximum frequency for different CGRA architectures. As can be seen, the GREEN variants reach higher frequency compared to other CGRAs. This is because of utilizing smaller logic components and/or reduced routing resources, both of which contribute to the critical path of the whole design. While XBioSip has the lowest frequency due to the large circuitry stems from the configurability of individual 2×2 multipliers on the LSBs up to 16-bit, in a 16×16 multiplier, X-CGRA and GP-CGRA present a moderate frequency, as they feature accuracy-configurability on a coarser grain (only two modes for each of adder/multiplier). Based on the Design Compiler synthesis reports, the inclusion of the 8-bit divider circuit has prolonged the critical path of X-CGRA, GP-CGRA, and baseline accurate architecture, compared to the GREEN variants. Following prior studies in the literature and to ensure a fair comparison for the power/energy calculations, the same synthesis frequency was set for all CGRAs (i.e., 300 MHz).

5.5 Conclusion

This chapter presented two approximate CGRA architectures, i.e., BioCare and GREEN which support parallelism of hybrid functionalities, for the first time, at the granularity of intra-PEs. The proposed CGRAs significantly improve performance (throughput) and energy efficiency, compared to accurate and state of the art counterparts. This is achieved through their light-weight SIMD/MIMD PEs. PEs, each of which enables function-versatility (ADD/MUL/DIV) and precision-adaptability (8-/16-/32-bit) by supporting a chain of approximations across the layers of abstraction (i.e., precision scaling on top of inexact multiplication and division). For the imprecise multiplication and division, the approximation strategy of SIMDive hybrid multiplier divider has been adopted, which facilitates the on-the-fly configuration of the SIMD/MIMD ALU to either of desired functionalities at run-time, without the need of reconfiguration.

The architecture-level evaluations have demonstrated that the proposed SIMD and MIMD configurations of our novel, (i.e., BioCare and GREEN, respectively) require up to 32% smaller resource footprint and 55% less energy, while enabling 49% higher speed than the accurate CGRA. Moreover, although augmenting the SISD version of the proposed CGRA, to also provide support for SIMD/MIMD execution modes has increased the complexity of SISD PEs, the maximum throughput of the designs is also increased up to 6.7×, compared to the accurate SISD counterpart.

Chapter 6
Application-Leveland Cross-Layer Approximations

Abstract This chapter introduces Plasticine, a cross-layer approximation framework designed to optimize multi-kernel programs for reconfigurable and embedded edge devices. While increasing throughput via SIMD and pipelined architectures, Plasticine tackles hardware constraints by leveraging application-level error tolerance. Existing approaches often approximate only single kernels with limited QoR analysis and overlook kernel-level contribution diversity. Plasticine addresses these gaps through a sensitivity analysis and a greedy-based heuristic that tailors approximation per kernel while meeting user-defined QoR. Its cross-layer chain spans application-, architecture-, and circuit-level , enabled by precision-adaptive SIMDive hybrid multiplier-dividers . Evaluated on bio-signal , image , and UAV tracking applications, Plasticine achieves significant area, latency, and Area-Delay Product (ADP) improvements over 32-bit fixed precision with negligible quality degradation.

Keywords Cross-layer approximation · FPGA · CGRA · SIMD · MIMD · High-throughput · Pipeline · Sustainability · Energy-efficiency · Area optimization · Error sensitivity analysis · Multi-objective heuristic · Design space exploration · Image processing · Bio-signal processing · ECG · EEG · Unmanned air vehicles · Neural networks · Classification

6.1 Introduction

On one hand, the rapid evolution of error-resilient programs intertwined with their quest for high throughput has motivated the use of SIMD or pipelined components in reconfigurable devices. On the other hand, to address the stringent resource constraints in the embedded devices (particularly in edge devices), designers are encouraged to exploit the error-resiliency of such applications. In fact, the cross-layer approximation paradigm has recently gained traction, the ultimate goal of which is to efficiently exploit approximation potentials across layers of abstraction. From circuit- to application-level, valuable studies have proposed various approximation techniques, albeit linked to key drawbacks: First, imprecise units are often substituted, merely in a single kernel of a multi-kernel application , with an end-to-end

Z. Ebrahimi and A. Kumar, *Cross-Layer Approximation and In-Network Acceleration*,
https://doi.org/10.1007/978-3-032-21712-7_6

analysis in QoR and not in the gained performance. Second, state of the art strategies neglect the fact that each kernel contributes differently to the end-to-end QoR and performance metrics. Therefore, they lack in adopting a generic methodology for adjusting the approximation knobs to maximize performance gains for a user-defined quality constraint. Finally, multi-level techniques lack in being efficiently supported, from application-, to architecture-, to circuit-level , in a cohesive cross-layer hierarchy

Therefore, in this chapter, Plasticine is presented, which sets out to enable a cross-layer methodology for multi-kernel applications and to address the aforementioned challenges by efficiently utilizing the synergistic effects of a chain of techniques across layers of abstraction. Plasticine methodology incorporates an application sensitivity analysis followed by a greedy-based heuristic that tailors the precision at constituent kernels of the application by finding the most tolerable degree of approximations for each of consecutive kernels, while also satisfying the ultimate user-defined QoR. The chain of approximations in Plasticine is also effectively enabled in a cross-layer hierarchy, from application-, to architecture-, to circuit-level, through the plasticity of SIMDive multiplier-dividers , each supporting dynamic precision variability along with hybrid functionality. The end-to-end evaluations of Plasticine on three multi-kernel applications employed in bio-signal processing , image processing , and moving object tracking for UAVs demonstrate 41%-64%, 39%-62%, and 70%-86% improvements in area, latency, and ADP, respectively, over 32-bit fixed precision, with negligible loss in QoR. The novel contributions of this chapter are listed as follows:

- The efficiency of the proposed imprecise multipliers and dividers (LeAp, SIMDive, RAPID) is evaluated individually at the application level. This is achieved by replacing the accurate counterparts with the imprecise configurations in the source code of the applications. The investigated application domains include image-processing (Image Blending and Gaussian Image Smoothing) and NN-based classification (MNIST handwritten digit recognition) applications and evaluating the final performance-accuracy trade-off.
- An end-to-end error-sensitivity analysis is proposed that determines the *relation* between the end-to-end performance gain and QoR loss for individual kernels when approximated by various techniques. This analysis helps to properly adjust the approximation knobs in our cross-layer approximation methodology. The analysis, evaluated in different single- and multi-kernel applications, demonstrates that the kernels contribute differently to the QoR fluctuations and the gains in *different* performance metrics, for *each* of the approximation techniques.

- A cross-layer approximation methodology for multi-kernel applications is proposed. The methodology is based on a novel greedy-based heuristic, the goal of which is to adjust various approximation knobs across the layers of abstraction. The efficiency of the cross-layer strategy is also assessed in both FPGA and CGRA and on various multi-kernel applications from different domains of application (bio-signal and image-processing to classification). The results show that the Pareto/near-Pareto *mixed-precision* configurations generated by our heuristic not only enable various performance-accuracy trade-offs but also render higher gains over accurate or uniform-precision counterparts in different accuracy levels.

In the rest of this chapter, the end-to-end error-resiliency sensitivity analysis is first presented, which serves as a prerequisite for the efficient implementation of the cross-layer approximation methodology. Afterward, the novel greedy-based heuristic is presented, which adjusts the cross-layer approximation knobs in multi-kernel applications. The considered knobs in this chapter are inexact operations (multiplication and division) along with the precision of kernels. Finally, the application-level and cross-layer evaluations on multiple single- and multi-kernel applications are discussed. The experiments have been conducted on both FPGA and CGRA.

6.2 End-to-End Error-Resiliency Sensitivity Analysis for Multi-Kernel Applications

A sensitivity analysis is needed to first understand the consequence of adjusting multiple optimization knobs to the error resiliency of the approximated application. This analysis was motivated by our initial observations that not only do kernels contribute differently to each of the performance metrics (e.g., area and latency), but also their significance order might differ in the *gained performance* and *lost QoR*, after applying different approximation techniques. Particularly, the observations have shown that not only kernels show different error-resiliency to the *same* approximation technique, but also their significance order in error-resiliency might change for each technique. These observations are highly important, as the proper adjustment of multiple approximation knobs (i.e., adjusting the intensity of applying different approximation techniques) should be based on the *trade-off*, i.e., the relation between the end-to-end ΔPerformance-gain and ΔQoR.[1]

[1] This fact was was neglected in prior works such as [Tas+20, Jud+15]. They adjusted the knobs only based on ΔPerformance-gain or Δ QoR.

The sensitivity analysis's main goal is to measure the error-resiliency of *each* application kernel to various approximation techniques. Revealing the relation (i.e., trade-off) between the gained performance and QoR-loss for individual kernels is critical to guide us in designing an approximation strategy subject to different approximation techniques. The considered techniques in our proposed analysis are adopted from different abstraction layers (to be applied in a cross-layer hierarchy). This chapter focuses on precision scaling (application-level technique) on top of inexact multiplication and division (circuit-level). In the next chapter, our methodology is expanded to also include other techniques, i.e., loop perforation (application-level) and changing the period and sampling frequency of data (at the data-acquisition level). The intuition behind adopting the precision scaling is three-fold. First, it reduces the number of computations, and the application's latency can also be reduced due to the shortened propagation delay of operations in the critical path. Second, precision scaling positively contributes to reducing the memory footprint and its associated data movement energy. Third, errors arising from neglecting LSBs will not result in a noticeable decline in the final accuracy (detailed results on different studied applications are discussed in the following). It should also be noted that addition and subtraction operations (circuit-level technique) were not approximated in the analysis for two reasons. The first is the small area and energy of individual adder/subtractor versus multiplier and divider (recalling Figs. 2.5, 2.4, and 2.6). The second is motivated by the experiments conducted in previous studies such as [HJi+20], stating that JPEG compression and other matrix-multiplication-based applications show higher accuracy-fluctuation to the approximation of additions rather than multiplication. In fact, for the applications having an aggregation-based structure (i.e., with a lot of addition and multiplication operations), low error bias plays a pivotal role in the imprecise component. Nevertheless, the analysis of different inexact adders could be further investigated in future tracks.

The sensitivity analysis is carried out by separately measuring the ① sensitivity of ultimate application' QoR to approximation of each kernel individually, and ② end-to-end performance gains obtained from each technique. The analysis is applied by approximating with a *one kernel at a time* strategy. This means that at each step, only one kernel is subjected to specific approximation technique(s), while the rest of the kernels are accurate. The output of the sensitivity analysis is n lists, each containing a pair of $\{\Delta$ Performance, Δ QoR$\}$ for each of the kernels subjected to that technique. This pair thus demonstrates the end-to-end trade-off between performance-gain and QoR loss, when only that kernel is approximated. The outputs of the sensitivity analysis, i.e., end-to-end performance gain and QoR loss—when different techniques approximate each kernel—are given in different lists to the cross-layer approximation heuristic (discussed in Sect. 6.3).

The following approach is exploited for applying the precision scaling on the results of intermediate kernels of the multi-kernel application. First, for 8-bit precision, the intermediate results are again scaled back to the 8-bit, before using in the next kernel, through finding the maximum value and then scaling, accordingly. It is worth noting that overflow cases mostly happen in 8-bit and the cases are infrequent in 16-bit (specially for the JPEG application). In Pan-Tompkins application multiplying ECG inputs with filter coefficients could result in overflows in the output of intermediate kernels. For this purpose, the output of each kernel is collected and the maximum value (Qmax) from that is determined. The entire samples are then divided by Qmax and rounded, thus bringing all the values in the range of (-1, 1). These values are further scaled up to the target bit-precision before being used in the next kernels. This approach were also used in other literature works [PRS19b].

Customizations for CGRA: For a semi-ASIC-based implementation a CGRA, this analysis is customized from two perspectives: (1) The synthesis flow and performance measurements have changed from an LUT- to the gate-level ASIC-based implementation. (2) A resource-efficient implementation varies for an ASIC-based mapping, which is considered in the implementation of case study applications (e.g., FFT [YJL20]). It is also worth noting that a sensitivity analysis for a CGRA platform also helps to estimate the minimum required number of PEs for individual kernels. Minimizing the number of PEs can also positively contribute to reducing the overhead of reconfiguration/data movement between consecutive kernels by, e.g., concurrently executing two kernels on the same CGRA.

6.3 Cross-Layer Approximation Heuristic for Multi-Kernel Applications

The main goal of the cross-layer approximation methodology is to appropriately adjust the approximation knobs in application kernels in a way to maximize the performance gain while minimizing the end-to-end quality loss (or maintaining a user-defined accuracy threshold). In fact, the importance of such methodology becomes pronounced for multi-kernel applications or in the presence of multiple approximation knobs: A fast and efficient search for finding the best configuration of knobs prevents the need for an exhaustive DSE.

To this end, a greedy-based heuristic was proposed in [Ebr+21], the pseudo-code and flowchart of which are demonstrated in Algorithm 7.4.1 and Fig. 6.1, respectively. The reason behind adopting a greedy heuristic, rather than other methods such as simulated annealing, ILP or GA-based alternatives, is as follows. The greedy heuristics have lower complexity than ILP and GA-based and faster convergence to the near-optimal solutions than the simulated annealing approaches.These are highly important factors for runtime decision-making actions in dynamic scenarios. In fact, the proposed heuristic generates *near-optimal* mixed-precision configurations—in a short amount of time—that can maximize performance gain for different user-defined accuracy thresholds.

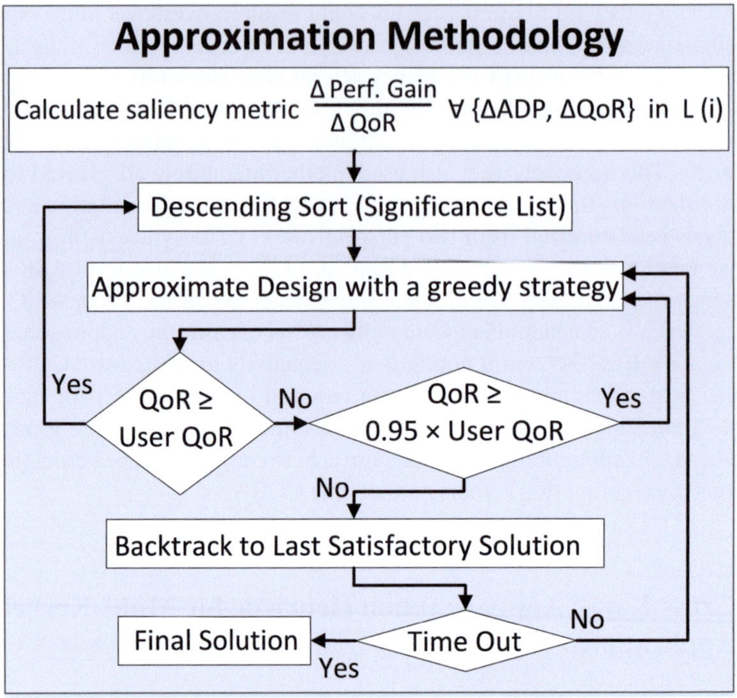

Fig. 6.1 Overall diagram for the cross-layer approximation methodology

The difference between our greedy heuristic and previous ones in the literature is as follows: In prior works, only the ΔQoR has been considered as the saliency metric for navigating the selection strategies at each step of their heuristic approaches. This neglects the effect of Δ Performance (here ADP), or in general, the relation between these two. Therefore, $\frac{\Delta ADP}{\Delta QoR}$ is proposed in the heuristic as the saliency measure, appropriately reflecting the potential end-to-end performance improvements over possible accuracy degradation when approximating a kernel. The inputs of our proposed design generation methodology are the user QoR constraint and the information gathered from the sensitivity analysis, resulting in four lists $L_1 - L_4$, are for approximating multiplication, division, or down-scaling the precision of the kernel to lower bit-widths (e.g., 16 or 8). Each list contains a pair of {Δ Performance-gain, Δ QoR} for each kernel, which demonstrates the end-to-end performance-QoR tradeoff of the application, when only that kernel is subjected to the target approximate technique and the rest of kernels are intact. As discussed, for performance gain, ADP improvement was opted for on FPGA platforms and throughput on CGRA. Nevertheless, other metrics can also be considered depending on the designer's objective.

The proposed heuristic, presented in Algorithm 1, works as follows: Initially, all application kernels are uniformly set to the highest precision, e.g., 32-bit with accurate operations. In the first step, the *Significance List* is formed by calculating the $\frac{\Delta ADP}{\Delta QoR}$ for each pair of {kernel, technique}. After merging the lists, it is sorted in a descending order, showing the significance order of techniques that resulted in providing higher performance improvement and less fluctuations in QoR, when considered individually. Then, the approximation is applied to kernels with an iterative approach: In each iteration, the heuristic applies an approximation that is placed on top of the significance list. The primary-applied techniques are projected to replace original multiplication and division with SIMDive versions, owing to their very low error metrics and high-performance gain potentials. Especially in the case of division, significant latency improvements are achieved by switching to a 32-bit SIMDive divider (46.1 ns in 32-bit accurate to 6.3 in 32-bit approximate, rather than down-scaling the precision with an accurate divider core (46.1–21.6 ns), referring to the circuit-level results in Table 4.8). The generated configuration is then appraised on miscellaneous samples to verify whether the end-to-end user-defined accuracy threshold is satisfied. Please note, error estimation formulation in multi-kernel programs—rather than applying a heuristic—is an open-research problem and not a claimed contribution of this book. Rather, the proposed $\frac{\Delta ADP}{\Delta QoR}$ factor enables a good vision to navigate toward *near-optimal* configurations in a short amount of time. Achieving credible results, the heuristic is evaluated on different tests for the application (e.g., ECG samples form MIT-BIH [Gol+00] for bio-signal processing domain or images from UAV123 [MSG16], VisDrone [Fan+20], or UAVid [Lyu+20] database for image-processing domain).

In the methodology, the freedom to explore *temporary* configurations was slightly increased, allowing negligible accuracy differences with the pre-defined threshold (1–5%). This freedom has enabled the heuristic to provide final QoR-satisfied solutions with also higher performance gain. The reason behind this is the following interesting observation: Sometimes, by combining approximated kernels, the accuracy slightly increases. Our profiling has revealed that it is partly due to the near-zero biased errors of SIMDive units that are able to neutralize each other in consecutive kernels, thereby *slightly* increasing approximation opportunities when combined. In this iterative approach, whenever the accuracy of the generated configuration crosses our threshold (with a 1–5% difference with the user-defined threshold), the heuristic backtracks to the prior user-satisfied configuration and follows the search by evaluating the next candidate, which has the highest priority. It is worth mentioning that the proposed cross-layer methodology for multi-kernel applications is agnostic to both the underlying architecture and the chain of techniques.

Algorithm 1: Cross-Layer Approximation Heuristic for Multi-Kernel Applications

Input: L1: {E2E Performance-Gain$_{xMUL}$, QoR-Loss} ∀ 32-bit Kernel
Input: L2: {E2E Performance-Gain$_{xDIV}$, QoR-Loss} ∀ 32-bit Kernel
Input: L3: {E2E Performance-Gain$_{Precision-Scale}$, QoR-Loss} ∀ Kernel, 16-bit Precision
Input: L4: {E2E Perf-Gain$_{Precision-Scale}$, QoR-Loss} ∀ Kernel, 8-bit Precision
Input: User-QoR-Constraint, Kernel-List
Output: Kernels [Architecture] /* *Precision and Mul/Div Structure* */

Significance-List = Array [] /* *Compute the effect of imprecise Mul, Div, and precision scaling, on each kernel, individually* */
for *i in Kernel-List* **do**
    Significance-List ← L1[i]$\frac{\Delta ADP}{\Delta QoR}(Mul)$
    Significance-List ← L2[i]$\frac{\Delta ADP}{\Delta QoR}(Div)$ /* *All dividers are approximated* */
    Significance-List ← L3[i]$\frac{\Delta ADP}{\Delta QoR}(16)$ /* *This kernel is scaled to 16-bit* */
    Significance-List ← L4[i]$\frac{\Delta ADP}{\Delta QoR}(8)$ /* *This kernel is scaled to 8-bit* */
end
Descending Sort (Significance-List);
while *(not timeout)* **do**
    for *i in Significance-List* **do**
        Design$_{\text{Approx}}$ = Kernels[Significance-List_i] /* *Approximate in descending order of* $\frac{\Delta ADP}{\Delta QoR}$ */
        Output-QoR = Evaluate (Design$_{\text{Approx}}$);
        if *Output-QoR* $\geq 0.95 \times$ *User-QoR-Const* **then** /* *Explore temporary configs* */
            if *Output-QoR* $\geq$ *User-QoR-Const* **then**
                Design_temp ← Design$_{\text{Approx}}$;
            end
            i ← i+1 ;
            go to 10 ;
        else
            Break;
        end
    end
end

6.4 Application-Level and Cross-Layer Results and Discussion

6.4.1 Experimental Setup

Application-level evaluation of individual multipliers: To further evaluate the efficacy of LeAp and SIMDive against state-of-the-art multipliers, they were deployed in multiple applications. To this end, behavioral implementations of the Image Blending application were developed, and accurate multiplication was replaced with different approximate alternatives. The final accuracy of the applications are evaluated on "*Miscellaneous*" images from USC-SIPI Database [Uni19].

We have also utilized approximate multipliers during the inference phase of a NN [CT16] for the classification of MNIST *Handwritten Digits* [LDe12] and *Fashion* [Xia+17] datasets. Two network configurations have been tested, i.e., two and three hidden layers. The total number of input nodes, hidden layers, and output layers are 28×28, 100, and 10, respectively. For both datasets, networks are trained with 50,000 images using Double Precision Floating-Point (FP64) numbers with a batch size of 10. Subsequently, during the inference phase, the networks are evaluated using FP64 and 8-bit fixed-point numbers. The results of these evaluations are detailed in Sect. 6.4.2, 6.4.2, and 6.4.2.

Partitioning and implementation of multi-kernel applications: The source-code of JPEG compression and K-means applications are adopted from AxBench [Yaz+17], while HCD and BSS are developed from scratch in C++. The base of JPEG and K-means applications are further optimized for a resource-efficient implementation, e.g., by transforming 2D-DCT computations to the butterfly-based 1D-DCT approach [Yaz+17, JOA13]. Moreover, for the Euclidean distance kernel used in the K-means application, the piecewise linear approximation method of Baptista [Wik17] was employed, as it is more straightforward and resource-efficient than general strategies for approximating the square root function [HJi+19, SF93, KL22]. Finally,the Pan-Tompkins algorithm is adopted from XBioSip [PRS19b]. These applications are partitioned into various *computational* kernels by excluding the critical portions (e.g., not related to memory or loop index calculation) or, e.g., instructions that are not implementable by the CGRA. Moreover, recalling Section 2.6, *zigzag* and *Huffman* kernels in JPEG compression implement a re-arrangement and encoding scheme. No approximations were applied in this process to remain in line with industrial standards. In addition, the *corner selection via non-maximum suppression*, also remained accurate in HCD case study, due to its moderately low resource footprint and the fact that it mainly comprises comparison operations.

FPGA mapping, synthesis, placement, and routing: The source code of the applications are synthesized using Vivado High-Level Synthesis (HLS) with DSPs being disabled to force the use of LUTs.[2] To efficiently reflect the optimizations of

[2] Using LUT-based soft multipliers for low bit-widths operations has also been recommended by e.g., Xilinx [Xil18a].

SIMDive in the final HDL design, a three-step approach was adopted to bridge this resource gap. In the first step, individual functions were coded for each size of accurate multiplication and division in the application. This process is also adopted for approximate SISD/SIMD designs (for SIMD configurations, the target operands are packed in a function to facilitate the replacement of SIMD modules with optimized SIMDive versions, later on in step 3. Afterward, using the HLS inline pragma, the compiler was forced to generate an independent HDL file for each function. Finally, the HDL description of the respective functions implementation—that realizes the target SISD or SIMD operations—are replaced by the HDL-optimized versions of SIMDive modules. For exact performance analysis, the HDL design of all accurate and approximate versions, generated by HLS, have been further passed to the downstream implementation phase, placed and routed on Virtex-7. HLS has two main advantages: First, it enables the flexibility to apply user-defined directives (when needed), more easily in a system-level implementation and shortens the process of generating various customized configurations. Second, it facilitates the high-level behavioral evaluation of approximate designs.

CGRA mapping, synthesis, placement, and routing: The applications are partitioned into various *computational* kernels (excluding instructions that are not implementable by the CGRA and the non-computational parts, e.g., related to memory or loop index calculation). Then, each of the application kernels are converted into a DFG—through an in-house C++ script as shown in Fig. 6.2—to be mapped on the CGRA. In the C++ coded DFG generator, each of the kernel operations is first assigned to a node (type *Struct*), which has a *function* field that reveals the type of the operation. Edges are also defined by another struct containing two fields, i.e., source and destination nodes. Afterwards, a DFG of type *Class* is defined that includes separate vectors for constructing the nodes and edges of the kernel. Then, through parsing the code, the nodes and their edges (the data dependencies between the operations) are extracted and added to the DFG, through calling separate functions in the C++ script. Moreover, in this phase, the number of load and store operations were also extracted for the energy calculations, presented in Sect. 6.4.3. Alternatively, this mapping process can also be carried through Low Level Virtual Machine (LLVM)-based tool-chains. However, it should be noted that the entire process cannot be carried through open-source CGRA frameworks such as Morpher [Wij+22], OpenCGRA [Tan+20], and CGRA-ME [And+21] due to their limitations, especially in mapping. For example these tools can only map a single unnested loop for each kernel, making the mapping process for a bigger program infeasible. In the DFG generation phase, no optimization technique was applied except loop transformation (loop unrolling and flattening of nested loops) and partial prediction (in which both operations of a conditional statement are assigned to different nodes/PEs).[3] Employing such techniques are orthogonal to our proposed architecture and approximation methodology and can be applied in tandem with the presented work. Afterward, though profiling

[3] Further elaborated *compiler-level* optimizations, such as Speculative Execution (which targets achieving the *optimal* Basic Block-BB size), were not adopted in the evaluations in this book for any of the CGRAs.

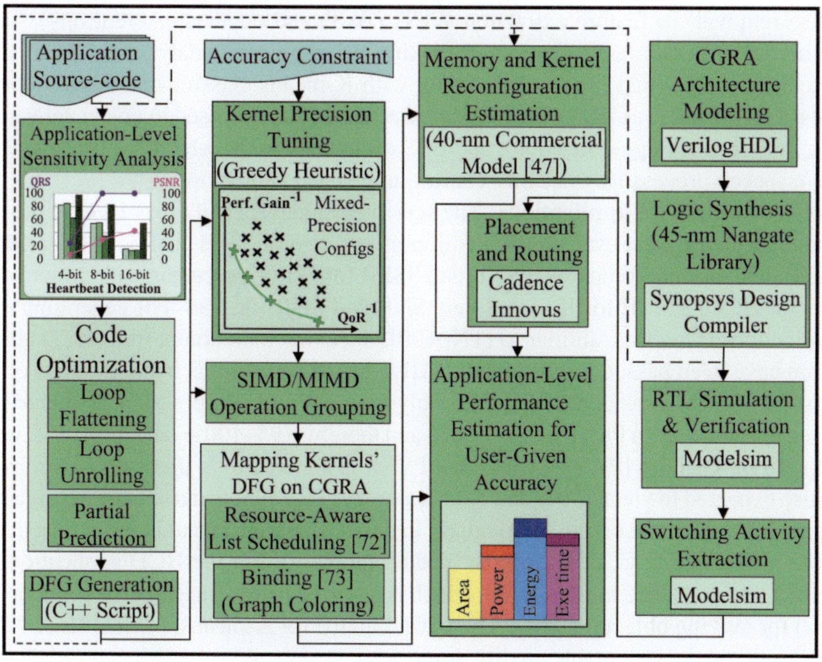

Fig. 6.2 Experimental setup, overall design, and approximation methodology in the GREEN CGRA

the output, common simple operations and/or SIMD/MIMD possible operations (i.e., without data-dependency) were grouped together considering both the architecture of the given SIMD/MIMD ALU and the sensitivity-analysis results that reveals whether the kernel can be approximated to a lower bit-width precision (the analysis results are discussed in the following). Finally, the DFGs are mapped onto the CGRA based on the widely used list scheduling algorithm (also adopted by HLS-based mapping approaches in the literature [PCL11]), which traverses the kernels' DFG nodes with a resource-aware approach [Mic94]. Dijkstra's algorithm [Dij22] has been employed to find the shortest path for the mapping (binding) of operations with data dependency that cannot be directly mapped to neighboring nodes. During the mapping, some PEs are inevitably used as routing nodes in cases that there is no direct connection for implementing two PEs with data dependency. Dijkstra's algorithm [Dij22] has been employed to find the shortest path.

Case study applications: Various single-kernel and multi-kernel applications from different domains were studied for the evaluation of the cross-layer approximation methodology.[4] Targeting single-kernel, have analyzed prevalent kernels shared by bio-signal applications, from pre-processing (band-pass FIR filter for

[4] The application-level (behavioral) evaluation of LeAp and SIMDive on the Gaussian smoothing filter, image blending, and MNIST NN are discussed in and.

noise removal) to feature extraction (FFT, DWT, and approximate entropy). Targeting multi-kernel tasks, the Pan-Tompkins algorithm was analyzed for heartbeat detection[5] (bio-signal processing), along with K-means classification, JPEG compression, and HCD as image processing case studies (widely used in object detection and motion tracking at the edge), and BSS of audio signals, which has broad usage from speech processing to quality control in industrial automation.

QoR analysis of applications: The sensitivity analysis and cross-layer approximation heuristic were evaluated with different application-level quality metrics. QRS and PSNR for Pan Tompkins, PSNR and SSIM for JPEG, percentage of correct vectors for HCD, RMSE for K-means, and SNR and SDR for BSS. For generating the motion/terrain vectors, similar to [TNo+20], the extracted corners from HCD algorithm have been passed through MATLAB simulations.[6] In this book, a PSNR of at least 28 dB was considered for ECG analysis and JPEG compression. It should be noted that although a PSNR of 11–19 dB, and thereby a 95–100% detection ratio, was allowed in XBioSip [PRS19b], a higher PSNR is adhered to in this work. This level is also expected to yield 100% detection accuracy [PRS19b] and to ensure acceptable signal strength in the output, particularly in cases where the signal is further used for other processing (e.g., detecting the type of anomaly). For the HCD application, a correct vector of 90% is considered, which is reported to be an acceptable confidence level for moving object tracking [TNo+20]. Finally, for K-means, the RMSE of 10% is considered the acceptable quality drop.

Benchmarks: For the Pan-Tompkins applications the ECG analysis are conducted on real-world MIT-BIH database [Gol+00] in MATLAB. For JPEG, K-Means, and HCD, end-to-end accuracy alterations were gauged on 100 miscellaneous images from different aerial imagery datasets [MSG16, Fan+20, Lyu+20]. Finally, 25 sets of 4-second audio samples from the Malfunctioning Industrial Machine Investigation and Inspection (MIMII) database [Pur+19] for BSS. This dataset represents the real-world data from industry machines, constructed by 26092 normal and abnormal operating acoustic data from four types of machines, i.e., valves, pumps, fans, and slide rails.

6.4.2 *Application-Level Evaluation of Individual Multipliers and Dividers*

Herein, the performance gain and quality loss of replacing imprecise multiplier and divider units (LeAp, RAPID, and SIMDive) are assessed in various applications.[7]

[5] It is not only an atomic task in heart disease diagnosis but is also employed in epilepsy/sleep apnea analysis, biometric authentication, etc.

[6] In [TNo+20], only DCT *adders* are approximated (DCT is used as a pre-processing compression step in studied UAV programs).

[7] The results are obtained without the use of sensitivity analysis and the precision-tuning heuristic.

(a) Original image **(b)** Accurate **(c)** LeAp-5 **(d)** MBM [HHS18a]

Fig. 6.3 PSNR (dB) of Image blending application computed with respect to the accurate multiplier-based filter: LeAp-5 PSNR=33.5, MBM [HHS18a] PSNR=32.6

(a) Noise-induced **(b)** Accurate **(c)** LeAp-5 **(d)** MBM [HHS18a]

Fig. 6.4 PSNR (dB) of Gaussian noise removal filter computed to the original noise-free image: **a** 20.5 **b** Accurate=28.7 **c** LeAp-5=28.4 **d** MBM [HHS18a]=27.9

Evaluation of LeAp Approximate Multiplier in image processing and NN applications: As discussed before, the efficacy of LeAp-5 was evaluated against the accurate and MBM [HHS18a] multipliers in the behavioral implementations of the Image Blending application. Examples of visual quality for both applications are illustrated in Fig. 6.5 and Fig. 6.6. Overall, the average PSNR value produced by LeAp-5 (32.19 dB) is better than MBM (31.23 dB). In another application (Gaussian Image Smoothing) LeAp also has surpassed MBM in PSNR (27.4 dB over 24.9 dB) (Figs. 6.3 and 6.4).

We have also utilized approximate multipliers during the inference phase of an NN [CT16] for classification of MNIST *Handwritten Digits* and *Fashion* datasets.

Table 6.1 shows the MNIST classification accuracy, area and energy gain by using approximate multipliers in 10,000 test images. As shown by the results, the error-resilience of NN makes the requirement for an accurate multiplier trivial for classification accuracy. In fact, for the MNIST *fashion* dataset, many of the quantization errors have been masked by approximate multipliers. For MNIST *handwritten digits* dataset, accurate and approximate multipliers-based networks produce almost similar classification accuracy. Interestingly, LeAp achieves precision of accurate multiplier while outperforming in terms of area and energy by 22% and 16% (also surpasses MBM in terms of energy). Note that referring to result in Table II, MBM will even have worse energy than accurate counterpart, provided that user exploits 16-bit multipliers in MNIST NN.

Table 6.1 Result of deploying LeAp and other multipliers in handwritten recognition application (MNIST NN)

MNIST dataset	No. of hidden layers	Nodes in each hidden layer	Inference accuracy %			
			Double Precision	**8-bit fixed precision**		
			Accurate	**Accurate**	**LeAp-5**	**MBM [HHS18a]**
Digits	2	100	97.09	96.67	96.67	96.62
Digits	3	100	96.56	96.23	96.22	96.17
Fashion	2	100	85.18	84.07	84.07	84.08
Fashion	3	100	85.29	84.26	84.27	84.39
Area (normalized to 8-bit accurate)			–	1	0.82	0.78
Energy (normalized to 8-bit accurate)			–	1	0.68	0.84

(a) Original image (b) Accurate (c) SIMDive=45.3 (d) MBM[HHS18a]=30.7

Fig. 6.5 Image blending application with various multipliers (PSNR values are w.r.t. the accurate multiplier-based filter)

Evaluation of SIMDive Approximate Multiplier-Divider (SISD mode) in image processing and NN applications: we have also evaluated the SISD mode of SIMDive multiplier and divider at application-level by deploying them in Image Blending and Gaussian Image Smoothing applications. The average PSNR value (in dB) produced by SIMDive-based application is 46.6, whereas it is limited to 32.1 for the MBM counterpart. Figure 6.5 presents visual qualities of two processed images. For the Gaussian Image Smoothing filter, the application was modified for two distinct modes: 1) only the approximate version of the divider is deployed. 2) Hybrid: Both multiplication and division are replaced with approximate alternatives. For the former case, SIMDive-based Gaussian filter is capable of producing Superior PSNR (24.5), over INZeD [HHS19] counterpart (20.9). Interestingly in the latter case, not only the PSNR value of hybrid SIMDive (23.3) surpasses the hybrid MBM/INZED (21.3), the values have not changed significantly compared to their *division only* approximation mode. This further motivates the deployment of hybrid SIMDive multiplier-divider. The visual quality of two processed images are shown in Fig. 6.6.

(a) Noise-induced=20.1 (b) Accurate=24.5 (c) SIMDive=24,23.3 (d) MBM/INZeD=21.3,20.5

Fig. 6.6 Gaussian noise removal filter (divider and hybrid multiplier-divider modes, PSNR values are w.r.t the original noise-free image)

Table 6.2 NN classification accuracy with accurate/approximate multiplier-divider

MNIST Dataset	No. of hidden layers	Nodes in each hidden layer	Inference accuracy %			
			Double precision	**8-bit fixed precision**		
			Accurate	Accurate	SIMDive	MBM/ INZeD [HHS18a, HHS19]
Digits	2	100	97.09	96.67	96.68	96.62
Digits	3	100	96.56	96.23	96.22	96.17
Fashion	2	100	85.18	84.07	84.09	84.08
Fashion	3	100	85.29	84.26	84.27	84.39
Area (normalized to 8-bit accurate)			–	1	0.78	0.74
Energy (normalized to 8-bit accurate)			–	1	0.62	0.72

We also evaluated the accuracy of MNIST and MNIST fashion classification by utilizing our proposed SIMDive-based approximate multiplier. The classification accuracy results are described in Table 6.2. Due to the inherent error resilience of NNs, many of the quantization induced errors have been partially healed for fashion dataset. Interestingly, SIMDive-based NN not only achieves same or higher classification accuracy compared to accurate and MBM/INZeD counterparts, it also outperforms accurate design and provides 22% and 38% improvement in terms of area and energy, respectively.

Evaluation of RAPID Pipelined Multiplier and Divider in Multi-Kernel Applications: the efficacy of RAPID over accurate and state of the art multipliers and dividers has been appraised in the end-to-end implementation of three multi-kernel applications, i.e., JPEG compression, heart-beat detection through Pan-Tompkins algorithm, and HCD. Figure 6.9 shows the obtained area, delay, and ADP improve-

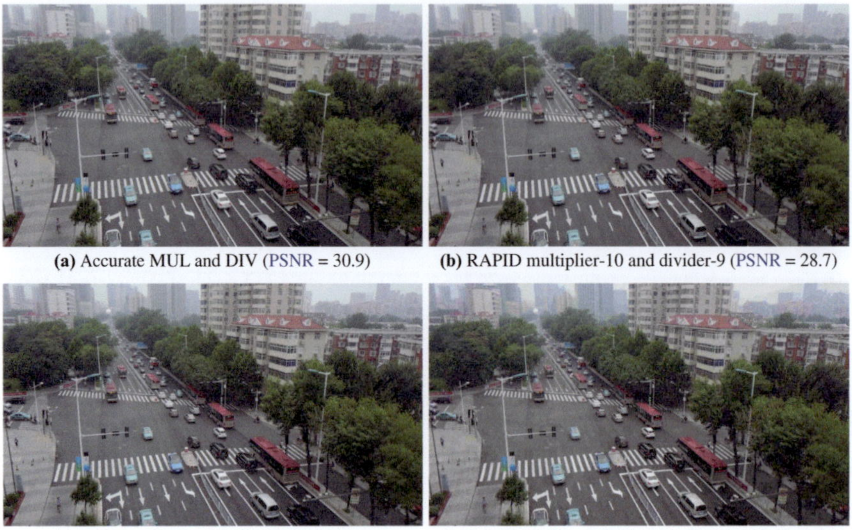

(a) Accurate MUL and DIV (PSNR = 30.9) (b) RAPID multiplier-10 and divider-9 (PSNR = 28.7)

(c) SIMDive MUL and DIV (PSNR = 29.3) (d) DRUM-6 MUL and AAXD-8/4 DIV (PSNR = 24.4)

Fig. 6.7 Comparison of JPEG compression on aerial images with accurate and different approximate multipliers and dividers (16-bit)

ments of approximate designs, when deployed in the three applications. The following inferences are highlighted on the efficacy of RAPID designs, after appraising the end-to-end performance gain (Fig. 6.7):

First, referring to Fig. 6.15 and Fig. 6.8 the quality drop by designs having a near-unbiased error characteristic (both RAPID and SIMDive) is smaller than when truncated counterparts are adopted. As discussed before, the biased error of truncated designs, i.e., DRUM and AAXD, results in accumulation of errors in consecutive kernels. In fact, our profiling has revealed that the near-zero biased errors of the RAPID multiplier and divider have been able to cancel out each other in consecutive operations/kernels and prevent a drastic error accumulation in the aggregation-based (i.e., having mostly addition/multiplication operations) structure of kernels. Second, the negligible accuracy loss after deployment of RAPID is partly due to the small ARE and PRE of the proposed multiplier and divider, as compared to high PRE in DRUM/AAXD, (up to 100% in hundreds of input cases). In fact, the analysis shows that a high PRE is also reflected as incorrect vectors in the HCD application (especially due to the presence of division in the last stage of the HCD algorithm). This also holds true for false positive heartbeat (QRS peaks): By utilizing both DRUM and AAXD the detection accuracy of heartbeat has dropped by $\sim$1%. Third, RAPID-configured applications also have better area- and ADP improvement, compared to both SIMDive and truncated counterparts (see Fig. 6.9). SIMDive on the other hand has marginally better end-to-end latency (by at most 3%), since its error-reduction circuitry is customized for performance-efficiency by directly configuring LUTs. Nonetheless, the difference of QoR and performance-gain between RAPID and SIMDive is not significant at application-level.

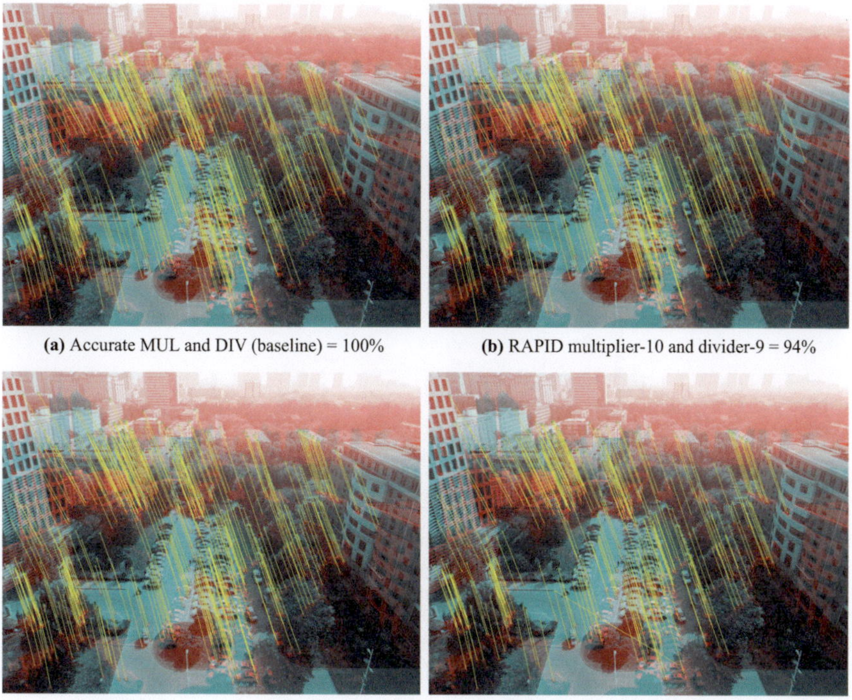

(a) Accurate MUL and DIV (baseline) = 100%

(b) RAPID multiplier-10 and divider-9 = 94%

(c) SIMDive MUL and DIV = 97%

(d) DRUM-6 MUL and AAXD-8/4 DIV = 83%

Fig. 6.8 Tracking via HCD: Changes in Harris score range also enables detection of new vectors in the threshold-based selection of tracking algorithm. Average of false positive vectors is < than 5%

To compare the throughput of applications in pipelined and non-pipelined configurations, the 2- and 4-stage pipelined versions of the RAPID multiplier and divider (along with their accurate counterparts) were also deployed in the applications. For a fair comparison, user-specified optimizations such as *function pipelining* pragmas (e.g., on matrix multiplication) were avoided, and the applications were implemented using a streaming approach. Figure 6.10 compares the end-to-end latency and throughput of both pipelined and non-pipelined configurations. Please note, the area difference by only replacing the non-pipelined RAPID multipliers and dividers with their pipelined versions is not significant (see Fig. 6.9). The formula for throughput is $Throughput = \frac{n}{(k+n-1)\times T_p}$, where n is the number of processed data samples, k is the number of pipeline stages, and T_p is the clock period of the pipelined architecture [Fly+95]. In computer architecture, when n>>k, the throughput is estimated as the inverse of the clock period (noting that this also holds true for image processing and bio-signal processing applications that are constantly fed with bulk of data). As can be seen in Fig. 6.10, the throughput of the applications increases after applying the pipelining, but this comes with the cost of increase in the end-to-end latency. Nevertheless, the latency overhead of converting RAPID_P2 configuration to RAPID_P4

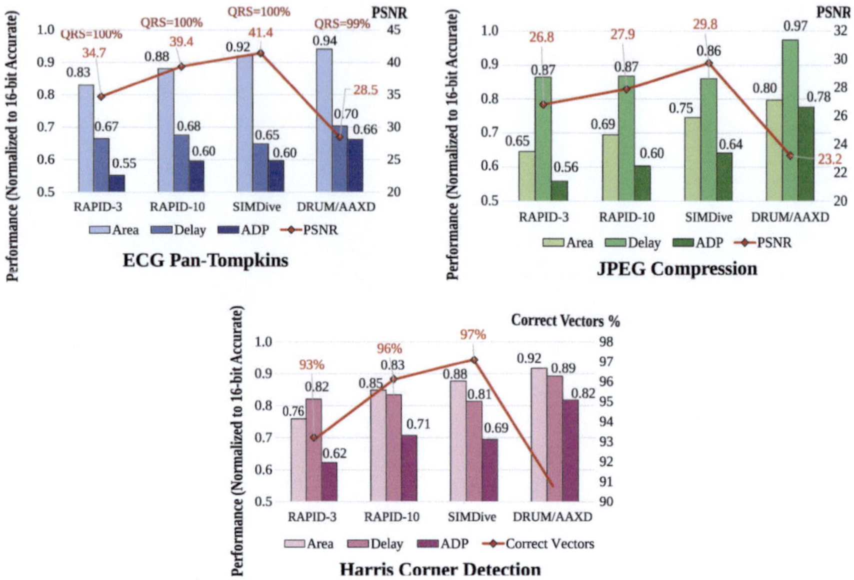

Fig. 6.9 End-to-end performance of applications utilizing RAPID multiplier and divider, compared to accurate and state of the art approximate designs

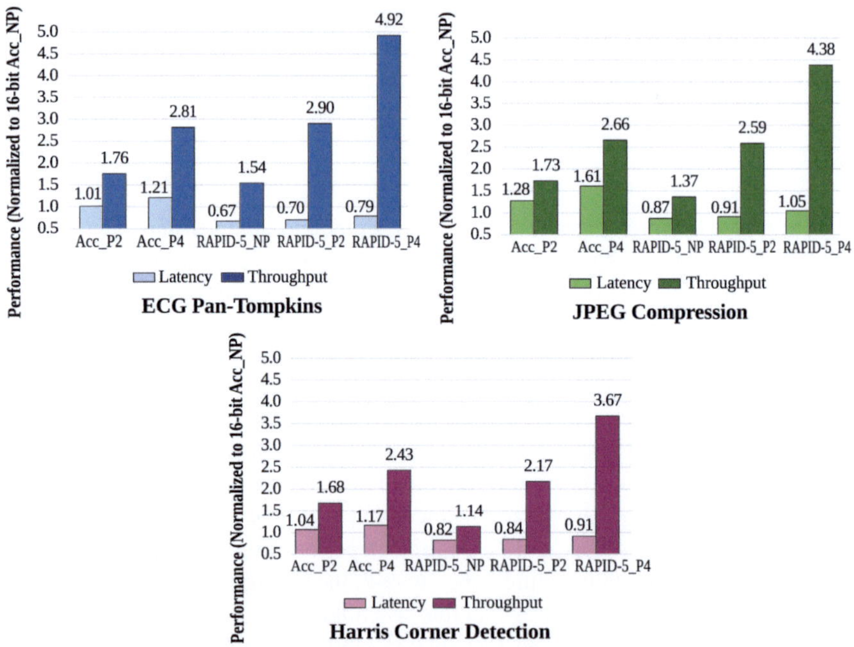

Fig. 6.10 The end-to-end latency and throughput of applications utilizing RAPID multiplier and divider, compared to accurate IP counterparts, in pipelined and non-pipelined format

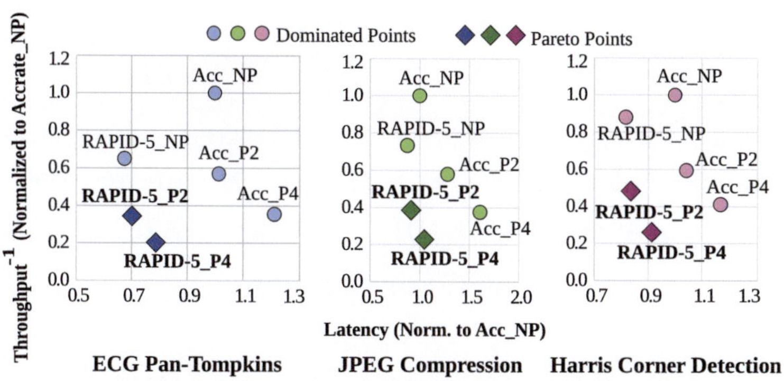

Fig. 6.11 Comparing the latency and throughput trade-off in applications exploiting pipelined and non-pipelined version of RAPID and accurate multiplier and divider

is less than of the overhead from converting Acc_P2 to Acc_P4. Similar observation also holds true for the improvement in the throughput. Moreover, application configurations having the RAPID_P2 reach smaller latency along with the higher throughput than when incorporating Acc_NP or Acc_P2. Finally, Fig. 6.11 illustrates the Pareto points in the trade-off between Latency and throughput for pipelined an non-pipelined architectures. As can be observed, RAPID_P2 and RAPID_P4 render the Pareto points having better trade-off in latency and throughput than other configurations.

6.4.3 Evaluating Cross-Layer Approximation Methodology

Evaluation of End-to-End Error-Resiliency Sensitivity Analysis
Herein, the results of the sensitivity analysis are presented, serving as a prerequisite for tuning the precision of different kernels that will be mapped onto a SIMD/MIMD reconfigurable FPGA or CGRA.

Results for FPGA: Fig. 6.12 summarizes the contribution of each kernel in the end-to-end LUT count, latency (Σ delay of consecutive kernels) of three multi-kernel applications, and ADP (as a representative of energy), when implemented in accurate mode. As can be seen, kernels contribute differently in each of performance metrics. This fact was not considered in most of the prior studies, when targeting approximating the application for a specific goal. For example, although most of the prior works have focused on approximating DCT kernel in the JPEG application (because of its multiplicative structure), our analysis shows that the *quantization* kernel contributes more significantly to the total application latency. In the sensitivity analysis, the effects of kernel precision scaling and imprecise multiplication and division on the QoR-performance trade-off are inspected. In this context, 32-bit integer precision is reported to reach a very high accuracy when implementing

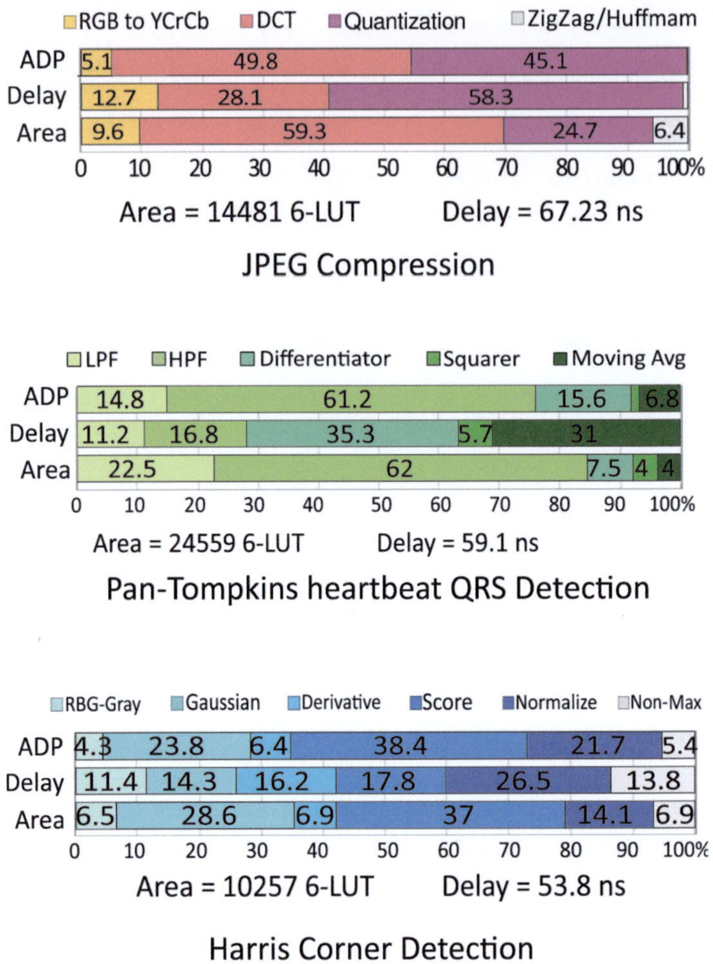

Fig. 6.12 Contribution of each kernel in performance metrics of application (accurate implementation)

the JPEG, HCD, and Pan-Tompkins applications [Sma18]. However, bounding the precision to 16- or 8-bit has also been reported to be satisfactory in recent studies [HHS19, HHS18a]. Therefore, the effect of downscaling to 16-/8-bit was analyzed as well. In this process, the SIMDive multiplier and divider, having 99.2% accuracy in all 32-bit kernels, were first deployed. Afterward, precision scaling was applied to one kernel at a time while others remained at 32-bit (with the intermediate results of each kernel scaled before being used in the next kernel). The obtained *end-to-end* performance improvements and their respective QoR for each of these configurations have been reported in Fig. 6.13. This figure, therefore, exhibits the sensitivity of kernels, individually, to approximations from different layers of abstraction (pre-

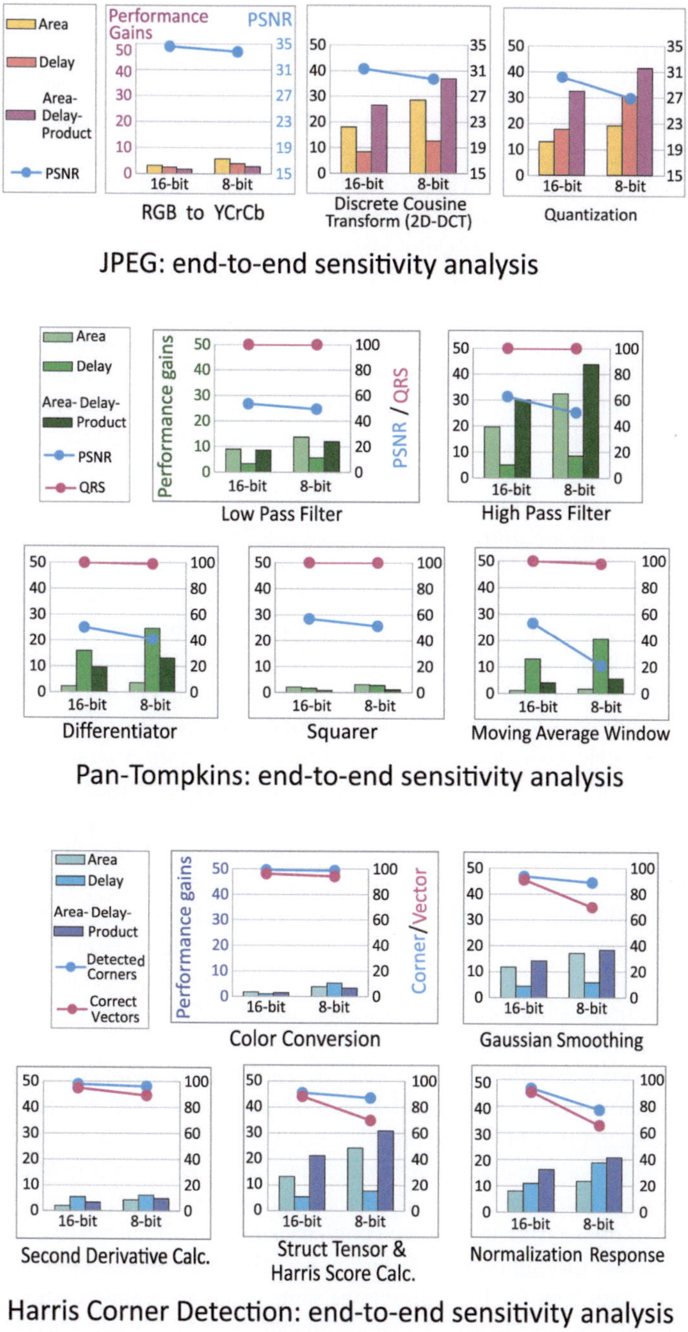

Fig. 6.13 Sensitivity analysis to cross-layer approximation techniques in three applications: The trade-off between end-to-end performance gains and QoR degradation after scaling down the precision of *one* kernel having SIMDive multiplier/divider, while the rest are 32-bit accurate (32-bit integer with accurate operations). The area is in terms of 6-LUT and delays in terms of nano-second.

cision scaling atop imprecise multiplication and division). The key observations of sensitivity analysis are pin-pointed in the following:

Deploying SIMDive marginally affects the accuracy: It was observed that replacing accurate multiplier and divider with SIMDive—having 99.2% accuracy—in all stages has marginally affected the final QoR, even when the precision of kernels is tuned at 16-bit. Fluctuations in the visual quality of JPEG compressed image and the number of detected corners and correct vectors in HCD were also negligible. Part (b) in each of Figs. 6.15 and 6.16 exhibit visual examples of such negligible changes by the deployment of SIMDive. The PSNR of 32-bit SIMDive based ECG is also 45.8 (the trade-off between QoR and ADP improvement in all configurations are shown in Fig. 6.19). This negligible accuracy loss is due to the small error metric of SIMDive, i.e., ARE and PRE. In fact, the analysis of HCD application has shown that more false positive or false negative vector cases will occur, provided that the approximate unit has a high peak error, i.e., PRE. For example, the AAXD divider has a PRE of 100%, as shown in Tables 4.7 and 4.8). The same also holds true for the heartbeat (QRS) peaks, especially considering that the 100% division error occurs at the last ECG stage. Second, our profiling has revealed that the near-zero biased errors of our multiplier and divider have been able to cancel out each other in consecutive kernels and prevent a drastic error accumulation in kernels' aggregation-based (mainly addition/multiplication operations) structure. Based on this observation, SIMDive was deployed in all kernels, thereby reducing the size of the design space to only changing the precision of kernels. In fact, our analysis exhibits that even 4-coefficient-based SIMDive results in 100% detection in a heartbeat or vectors generated by Harris corners. Nonetheless, the accuracy of multiplication/division was increased (at the negligible cost of a few additional LUTs), creating an opportunity for adaptable precision scaling.

Applications show more sensitivity to approximation of addition: It was observed that truncating the output of addition/subtraction in kernels (beyond a few bits) significantly affects accuracy; therefore, addition is kept accurate in the depicted figure. This happened, especially in *differentiator* or *moving average window* stages of the Pan-Tompkins application and *corner response calculation* stage of HCD. As discussed earlier, such observations are also endorsed by, e.g., [HJi+20] for matrix multiplication-based applications.

Kernels contribute differently to each of performance metrics: Fig. 6.13 also reveals that kernels exhibit different behavior w.r.t the gains from the same approximation. For example, when targeting LUT saving, approximation of the High-Pass Filter (HPF) (second kernel) is more beneficial in ECG analysis, while when improving the heartbeat detection latency is the goal, *differentiator* should be approximated (because of its division, acting as speed bottleneck operation). Interestingly, the same figure also shows that although DCT has higher ADP than *quantization* in the original accurate configuration, by applying the same precision scaling on both kernels, more energy saving is achieved via approximation of *quantization*. This is because, beside LUT, latency of *quantization* also significantly reduces (due to its divider). This further corroborates our strategy that the approximation of kernels should be adjusted based on the *gained* improvement, not based on their primary appearance order in the non-approximated version.

Kernels significance order in the QoR loss can differ from its order in the gained performance: While applying the same approximation results in a specific order for the gained performance, the same or inverse order will not necessarily appear in QoR fluctuations. For example, referring to Fig. 6.13, amongst Pan-Tompkins kernels, *HPF* enables the highest LUT saving and affects the ultimate PSNR, the least. However, as shown in this figure, the *Harris score* kernel enables the highest LUT saving and changes the final accuracy more than other down-scaled kernels. Hence, our insights endorse that: First, the *relation* between ΔPerformance and ΔQoR should be considered. Second, applying the most aggressive approximation on early stages (which is the adopted strategy in XBioSiP [PRS19b]) not only neglects order of stages w.r.t their QoR sensitivity but also disregards their order in the obtained performance improvement.

Conclusion: In short, the aforementioned insights corroborate our strategy, which states that for the approximation of multi-kernel applications, one should consider the *relation* between the *gained performance* and *QoR loss*. Moreover, provided that the multi-kernel application has an aggregation-based structure, adopting imprecise operations with minimally biased errors in all kernels can create the opportunity for applying a run-time adaptable precision scaling (efficiently supported herein, thorough on-the-fly configurable SIMDive units). The outputs of this sensitivity analysis, i.e., end-to-end performance gain and QoR loss, when each kernel is approximated by different techniques (imprecise multiplication, division, and scaled down to 16- or 8-bit) are given in four lists to the cross-layer approximation methodology, discussed in Sect. 6.3.

Results for CGRA: In the sensitivity analysis customized for CGRA, the effect of inexact multiplication-division and precision scaling on the end-to-end accuracy and performance metrics of the applications is explored. A 16-bit baseline precision was considered, as it has been shown to deliver acceptable accuracy for the case study applications. Therefore, precision scaling was applied to 8- and 4-bit configurations. Figure 6.14 depicts the results of the sensitivity analysis on a near-ASIC implementation exploited in a CGRA platform. The summary of the results is discussed below:

Substituting accurate multiplier-divider with SIMDive: Deploying SIMDive multiplier-divider in the kernels has marginally affected the QoR in most of the applications. For example, as shown in Fig. 6.14, the PSNR of 16-bit SIMDive-based Pan-Tompkins applications is 42.3, and the QRS detection is maintained at 100%. Such a good PSNR is endowed to the near-zero biased error of SIMDive units (which can cancel out each other) when exploited in the aggregation-based structure (i.e., mainly addition/multiplication operations) of the kernels [Ebr+21]. Based on this observation, a homogeneous structure of PEs was exploited, all utilizing the SIMDive units (rather than collecting separate versions of accurate multipliers and dividers, as adopted in other approximate CGRA works in the literature [Akb+20, OAk+18, Akb+18, Bra+18]). In contrast, the approximation of the addition functions can adversely affect the accuracy, especially when applied in kernels such

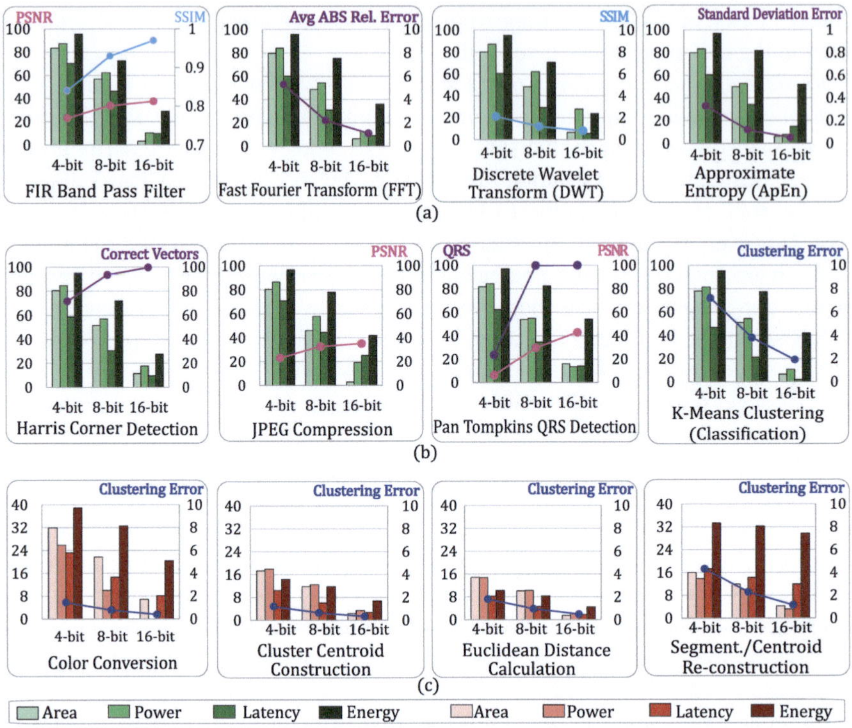

Fig. 6.14 (Top) Sensitivity analysis of seven single-kernel applications to SIMDive multiplication and division and precision scaling (Reference: 16-bit kernels with accurate multiplication-division). (Down) A detailed example of the sensitivity of a multi-kernel case study (K-means application)

as *differentiator* or *moving average window* in Pan-Tompkins algorithm or *corner response calculation* in HCD. This observation justifies our statement that considering the small contribution of adder in the total energy of the PE (recalling Figure 2.6), it is better to preserve the addition function accurately and instead focused on an adaptable precision scaling strategy.

Down-scaling precision from 16- to 8-, and 4-bit: Analyzing the real-world ECG and EEG signals from MIT-BIH and CHB-MIT databases [Gol+00] shows that the samples are unevenly distributed in the range of 16-bit: 94.8% of ECG and 92.2% of EEG input samples can be trimmed to be fitted into 8-bit, respectively, without affecting final QoR significantly, when passed through the calculations. In fact, even the uniform 8-bit precision with accurate operations provides nearly 100% heartbeat QRS detection. Moreover, the fluctuations in the image PSNR and the number of detected corners/correct vectors in HCD, after down-scaling the precision to 8-bit, were also less than 1% and 7%, respectively. However, setting the uniform precision of 4-bit for *all* kernels has resulted in a noticeable accuracy drop for some applications, especially the ones that included more than one division operation. For example, as shown in Fig. 6.14, the QRS detection ratio is drastically dropped to

$\sim$24% in Pan-Tompkins algorithm or RMSE in K-means application goes beyond 16%. Such observations demonstrate that a mixed-precision strategy is better suited for approximating multi-kernel applications.

Kernels/operations contribute differently to the QoR changes and the gains in different performance metrics: As can be observed in Fig. 6.14, applying approximation may result in different orders for Δ*performance gain* and ΔQoR. For example, as can be seen in the K-means case study, the approximation of *color conversion* is more beneficial when area improvement is the goal. In contrast, *cluster centroid construction* is more error-resilient to approximation. In fact, the order of importance of kernels may differ, even for different performance metrics. Such observations are also corroborated for other applications when approximated by different approximation techniques, e.g., LSB truncation of addition and multiplication functions [PRS19b, Ebr+21].

Conclusion: In short, the insights mentioned above again corroborate our earlier findings and statements that a mixed-precision strategy can enable different energy-accuracy trade-offs while reducing, e.g., the required for executing the application with acceptable accuracy. This is achievable by our approximation heuristic and on-the-fly configurable SIMD/MIMD GREEN CGRA architecture discussed in 5.3.

Evaluation of Cross-Layer Approximation Heuristic in Multi-Kernel Applications Herein, the efficiency of the cross-layer approximation heuristic is assessed in various multi-kernel applications with respect to the end-to-end performance gain and quality loss of the application. Concluded by the sensitivity analysis, the imprecise SIMDive multiplier-divider was adopted as the circuit-level technique and precision scaling as the application-level technique. The main reasons for adopting SIMDive in the case study applications is its ability to 1) be configured for SIMD mode (beneficial for parallelizable tasks such as matrix multiplication) and 2) efficiently reflect the gains from precision scaling in a cross-layer hierarchy.

Evaluation of cross-layer heuristic on an FPGA-based implementation

Approximate mixed-precision configurations versus accurate and uniform counterparts: Fig. 6.19 details the obtained ADP improvement for different kernel configurations and distinguishes the mixed-precision configurations found by the heuristic. For packing SIMD operations inside precision reduced kernels, dependency of multiplication and/or division operations has been considered (referring to Figures 2.7, 2.8, and 2.9). Empirical observations from sensitivity analysis such as adopting the approximate multiplier and divider in all kernels and maintaining accurate structure of *addition* operation, has pruned the dominated points and efficiently reduced the size of design space that is evaluated in our case studies: As detailed in Table 6.3, simulating 243 possible ECG scenarios in MATLAB (each of five kernels can be adjusted to 8-, 16-, or 32-bit) takes a runtime of 11.7 h, when analyzing 20k real-world ECG samples of MIT-BIH database, sampled at the frequency 200 Hz. Run-times of 27 scenarios in JPEG and 243 scenarios in HCD (HLS generated corners processed in MATLAB based code to produce terrain vectors) also take 5.2 and 62.6 h over 100 image samples, respectively. The following inferences are highlighted on the functioning and efficacy of our cross-layer strategy.

(a) Uniform 32-bit, accurate MUL/DIV=32.7

(b) Uniform 32-bit, approximate MUL/DIV=31.8

(c) Uniform 16-bit, approximate MUL/DIV=30.3

(d) Mixed-precision, approx. MUL/DIV=29.1

Fig. 6.15 JPEG compression for aerial images: The PSNR fluctuations are still tolerable by applying precision scaling and SIMDive MUL/DIV approximation.

(a) Uniform 32-bit, accurate MUL/DIV (100%)

(b) Uniform 32-bit, approximate MUL/DIV=97.8%

(c) Uniform16-bit, approximate MUL/DIV=94.4%

(d) Mixed-precision, approx. MUL/DIV=92.1%

Fig. 6.16 Tracking via HCD technique: The number of detected corners and vectors marginally changes with approximation (changes in Harris score range also enables detection of new vectors in the threshold-based selection of tracking algorithm. Average of false positive vectors is less than 4%)

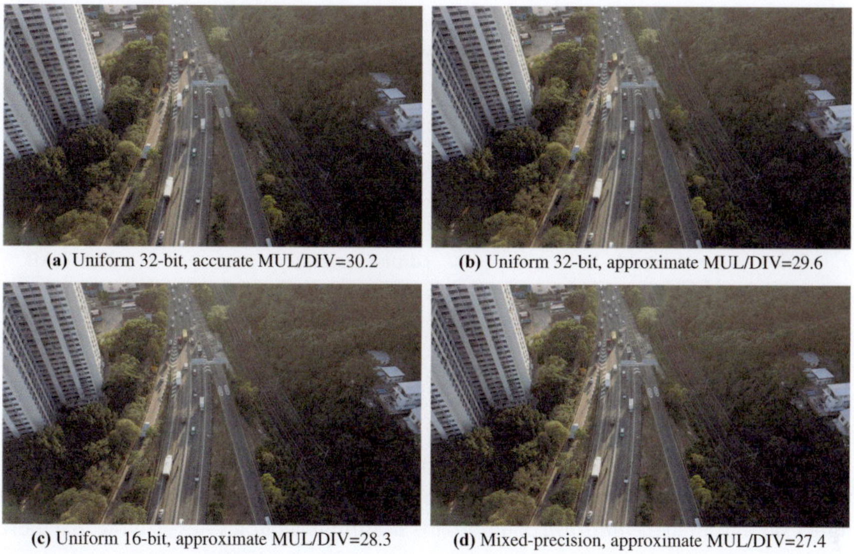

(a) Uniform 32-bit, accurate MUL/DIV=30.2

(b) Uniform 32-bit, approximate MUL/DIV=29.6

(c) Uniform 16-bit, approximate MUL/DIV=28.3

(d) Mixed-precision, approximate MUL/DIV=27.4

Fig. 6.17 PSNR in JPEG application: Visual quality of image is gracefully maintained after approximations

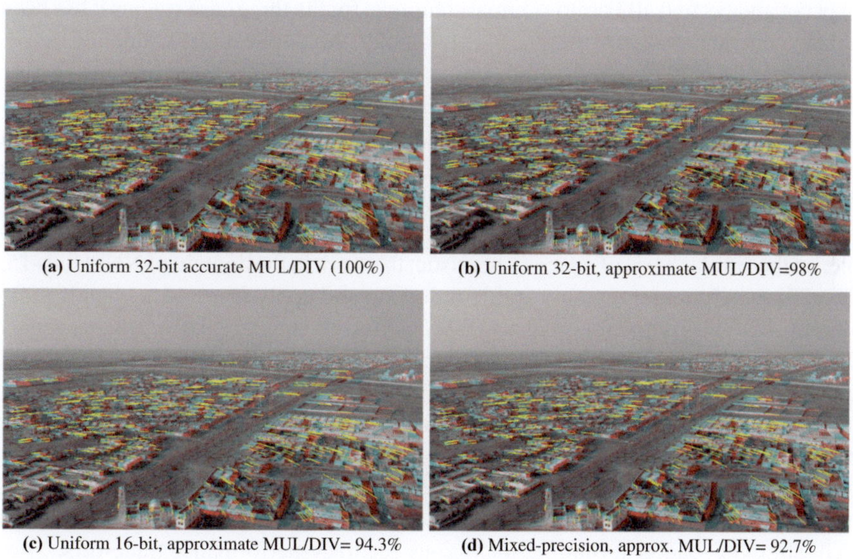

(a) Uniform 32-bit accurate MUL/DIV (100%)

(b) Uniform 32-bit, approximate MUL/DIV=98%

(c) Uniform 16-bit, approximate MUL/DIV= 94.3%

(d) Mixed-precision, approx. MUL/DIV= 92.7%

Fig. 6.18 Corners and vector detected via HCD application: As can be seen even with approximation the direction of the movement has been found by many vectors

Table 6.3 Simulation setup and exhaustive run-time for three applications (multipliers and dividers in the application kernels are replaced by SIMDive)

	Pan Tompkins (ECG)	JPEG compression		**Harris corner detection**	
# Configs	Exhaustive time	# Configs	Exhaustive time	# Configs	**Exhaustive time**
Five kernels (8/16/32-bit) $3^5 = 243$	11.7 h (20k real-world ECG samples)	3^3 = 27 configs	5.2 h in HLS (100 drone images)	$3^5 = 243$	62 h: HLS (corners) + MATLAB (vectors) (100 drone images)

First, it can be deduced from Fig. 6.19 that not only new resource-accuracy trade-off levels are generated by *mixed-precision* configurations but also they can enable higher savings than the *uniform* counterparts in similar accuracy-level. Note that there is not only one unique but a set of configurations with permissible resource-accuracy trade-offs (few visual examples are shown in Fig. 6.15, 6.17, 6.18, and 6.19).

Second, although the proposed design generation methodology is a heuristic and may not always find all Pareto-optimal solutions, for the three case studies it has been able to generate most of Pareto/near-Pareto points for applications within the first $\sim$ 10-30% of their execution time (the points generated in first 20% of exhaustive time are shown in green stars in Fig. 6.19). This asserts the effectiveness of the heuristic and its notable saving in exploration time, especially when brute-force exploration is not time-wise tractable for a larger design space. Note, Fig. 6.19, shows how the heuristic proceeds toward the Pareto points (when the user-QoR threshold is 70%).

Third, $\frac{\Delta ADP}{\Delta QOR}$ is customizable w.r.t the design constraints and objectives. For example, on the similar basis of a greedy heuristic, the work in the domain of NNs [Jud+15, Tas+20] mostly target memory traffic reduction and apply *layer-wise* quantization. Herein, ADP was opted for, as in an FPGA platform a smaller ADP can better reflect the resource savings. Nevertheless, other performance metrics can also be explored based on designer's goal.

Finally, this flexibility provided by our adaptable approximation strategy also can be exploited in a run-time energy-management technique: The bitstream of arbitrary heuristic-generated configurations or Pareto-optimal configurations (provided that exhaustive search is feasible) can also be stored in the FPGA SPI/BPI flash memories [Xil18b] to be loaded at runtime for various accuracy-performance trade-offs.

End-to-end performance gains of heuristic-generated configurations: Herein, the end-to-end QoR-performance trade-off for five approaches are evaluated on Pan-Tompkins, JPEG compression, and HCD applications. Approximation approaches comprises the two proposed, i.e., uniform 16-bit and mixed-precision SIMDive based configurations, found by the proposed $\frac{\Delta ADP}{\Delta QOR}$ based heuristic. These approaches are compared against two baselines, i.e., uniform 32- and 16-bit accurate counterparts and mixed-precision SIMDive based configurations, found by ΔQOR based heuristic

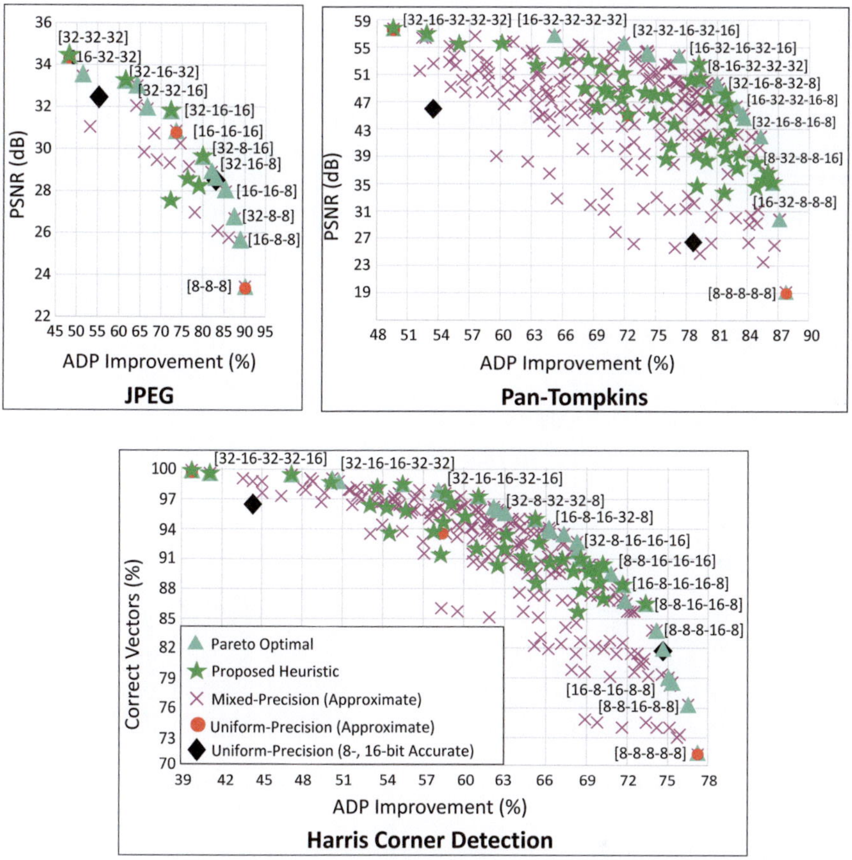

Fig. 6.19 *Approximate, mixed-precision* configs generate accuracy-performance trade-offs with similar/higher gains than *accurate-uniform*. The heuristic also finds many of Pareto/near-Pareto points. Configs found in 20% of exhaustive time, are distinguished by green stars (QoR thresholds: JPEG 28 dB, ECG 35 dB, HCD 90%)

(adopted in [Tas+20, Jud+15]). For these quality thresholds discussed in Sect. 6.4.1, the following configurations are found by the proposed heuristic: [8-8-16-8-16] for five-kernel ECG Pan-Tompkins, [8-8-16] for three-kernel JPEG Compression, and [8-8-8-16-16] for five-kernel HCD. Accordingly, the adopted SISD/SIMD structure of SIMDive for the proposed kernel configurations is detailed in Table 6.4.[8] The end-to-end performance gain from such heuristic-generated configurations are evaluated against the baselines, in Fig. 6.20. The results detailed in this figure, demonstrates that not only the mixed-precision configurations enable higher improvement over the uniform 16-bit precision counterpart but also the proposed $\frac{\Delta ADP}{\Delta QOR}$-based heuris-

[8] The reported gains are w.r.t the *computations*, without considering the overhead of packing.

Table 6.4 Kernels' SISD/SIMD structure in three application, w.r.t the final approximate configuration

Pan tompkins (ECG)		JPEG compression		Harris corner detection	
Kernel name	Configuration	Kernel name	Configuration	Kernel name	Configuration
Low-pass filter	SIMD 8-16 bit	RGB to YCbCr	SIMD 8-16 bit	RGB to Grayscale	SIMD 8-16 bit
High-pass filter	SIMD 8-16 bit	2D-DCT	SIMD 8-16 bit	Gaussian Smoothing	SIMD 8-16 bit
Differentiator	SISD 16 bit			Derivative Sobel	SIMD 8-16 bit
Squarer	SIMD 8-16 bit			Tensor & Score Response	SISD 16 bit
Moving avg filter	SISD 16 bit	Quantization	SISD 16 bit	Normalization	SISD 16 bit

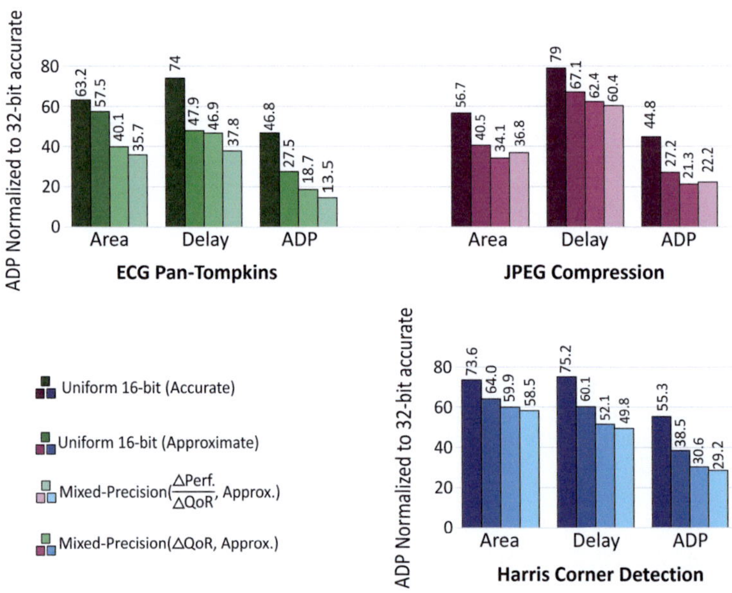

Fig. 6.20 End-to-end performance gains of different strategies, compared to 32-bit accurate (uniform) configuration: Overall, mixed-precision configurations, generated by the proposed $\frac{\Delta ADP}{\Delta QOR}$-based heuristic achieves higher gain than the ΔQoR-based heuristics [Tas+20, Jud+15]

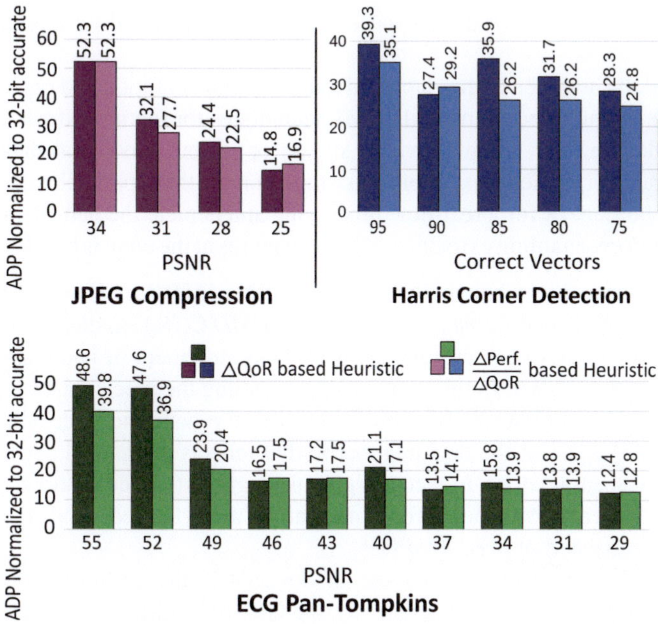

Fig. 6.21 Comparing $\frac{\Delta ADP}{\Delta QOR}$-based versus ΔQoR based heuristic in various QoR levels: The former approach has generated configurations with higher ADP improvement, when the accuracy threshold is high

tic outperforms the ΔQoR-based approach [Tas+20, Jud+15], in ECG and HCD applications.

Our proposed $\frac{\Delta ADP}{\Delta QOR}$-based greedy heuristic is also appraised against the baseline (which only considers ΔQOR as the saliency metric) are appraised on various accuracy thresholds in all applications, the results of which is presented in Fig. 6.21. It should be noted that the functionality and thereby, the quality of the final configurations found by these heuristic approaches depends on different factors, including kernel structure, exploration time, accuracy-performance trade-off of the approximate components, etc. Nevertheless, as detailed in Fig. 6.21, it was observed that the proposed heuristic can find solutions with higher performance improvement when the accuracy threshold is moderate to relatively high. On the contrary, when the ultimate quality threshold is relaxed, ΔQoR-based heuristic takes more steps with smaller quality changes (which also increased the heuristic exploration time) and gradually approaches to a solution with slightly higher gains, up to 8%. It should be highlighted that, to fairly assess the quality of the solutions found by the ΔQoR-based heuristic against those found by the proposed approach (shown in Fig. 6.21), the same 1-5% freedom was enabled for both approaches, and the heuristic was started after the design space had been efficiently pruned by the proposed sensitivity analysis. Comparing the search complexity of various heuristics such as [LJG17], for which the original size of design space is different, is planned for a future track of this book.

Discussion: It should be highlighted that although more complicated object tracking or heart arrhythmia programs utilize ML techniques for feature extraction, continuously offloading the complete video stream or patient bio-signal data to the insecure/untrustworthy network, or process/store it on the third-party cloud can pose performance bottleneck for a real-time processing and deplete the battery in a short interval. Therefore, enabling extraction of some features at the edge is highly desired.

Evaluation of cross-layer heuristic on the GREEN SIMD/MIMD CGRA

Precision tuning of kernels for mapping on SIMD/MIMD CGRA: as discussed earlier, our greedy heuristic has been customized w.r.t the requirement of applications that are executed on the CGRA. In this context, down-scaling the precision of the kernel has been from 16- to either 8- or 4-bit. Moreover, the heuristic is customized to consider the $\frac{\Delta \text{Throughput}}{\Delta \text{QOR}}$ as the deciding metric that reflects the gained performance over a possible accuracy loss. In fact, a higher throughput (gained by down-scaling the precision) better reflects the performance gain in a pre-fabricated SIMD/MIMD CGRA that is consisted of *fixed-area* PEs.

For the quality thresholds discussed in Sect. 6.4.1, the following precision-configurations are found by the heuristic for the GREEN CGRA: [4-8-4-16-16] for the HCD having five-kernels, [4-4-8-4-16] for the five-kernel ECG Pan-Tompkins QRS detection, [4-4-8-16] for the four-kernel K-means, and [4-8-16] for the three-kernel JPEG Compression. Accordingly, the adopted SISD/SIMD/MIMD structure of kernels for the proposed kernel configurations is detailed in Table 6.5. It should be noted that for packing the SIMD/MIMD operations inside the precision reduced kernels, the dependency of multiplication and/or division operations (referring to Figure 2.10) has been taken into account.

We have adopted the one-kernel-at-a-time strategy of [Duc+17], suited for the processing of bio-signals, as these applications have brief period of sampling followed by a relatively large idle interval. In this strategy, the output of intermediate kernels are stored in the on-chip memory—which are directly accessible by PEs—and fetched when PEs are configured to implement the next kernel (an 8×8 array suffices mapping the biggest sub-kernels of the case study applications).

B*asic* B*lock (BB) size and the number of used PEs for mapping the block*: The BB size is determined directly based on the result of the resource-aware list-scheduling/mapping, after determining the precision of PEs in SIMD/MIMD modes (also shown in Table 6.5).[9]

The pair of {*BB size, the number of used PEs for mapping the block*}, for different 8×8 architectures and application kernels, is reported in Table 6.6. As can be seen, for the kernels running at 4- or 8-bit precision, the SIMD and MIMD architectures have been able to implement larger BB size within the same or smaller number of PEs,

[9] Please note, BB size enables a trade-off between exploiting parallelism and minimizing communication overhead. Therefore, an optimal block size can vary depending on various factors, including the CGRA architecture, characteristics of the application kernel (e.g., data dependency), the optimization techniques, e.g., loop unrolling.

Table 6.5 Kernels' SISD/SIMD/MIMD structure in multi-kernel application, executed on the GREEN CGRA

Harris corner detection		**Pan-tompkins (ECG)**	
Kernel name	Configuration	Kernel name	Configuration
RGB to Grayscale	SIMD 4-bit	Low-Pass Filter	MIMD 4-bit
Gaussian Smoothing	MIMD 8-bit	High-Pass Filter	SIMD 4-bit
Derivative Sobel	MIMD 4-bit	Differentiator	MIMD 8-bit
Tensor & Score Response	SISD 16-bit	Squarer	SIMD 4-bit
Normalization	SISD 16-bit	Moving Avg Filter	SISD 16-bit
K-Means Clustering		**JPEG Compression**	
Kernel Name	**Configuration**	**Kernel Name**	**Configuration**
RGB to YCbCr	MIMD 4-bit	RGB to YCbCr	MIMD 4-bit
Cluster Cent. Const.	SIMD 4-bit	2D-DCT	MIMD 8-bit
Euclidean Distance Calc.	SIMD 8-bit		
Segmentation/Cluster Cent. Re-Construction	SISD 16-bit	Quantization	SISD 16-bit

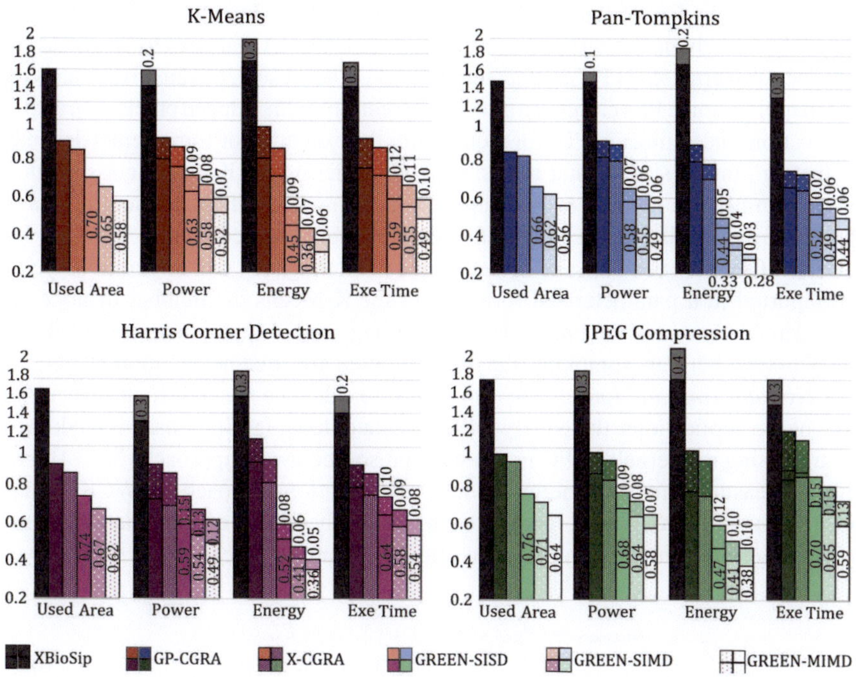

Fig. 6.22 Application-level comparison of CGRAs, by running multi-kernel applications

compared to the SISD architecture. This is due to packing more instructions (with lower precision) in a single PE, enabled by the SIMD/ MIMD modes. Therefore, in case of availability of resources, the user can map more BB instances onto the same CGRA array.

Utilization of the 2D array for the different kernels: Table 6.7 details the utilization of 8×8 array for different kernels and types of configuration (SISD, SIMD, and MIMD). Kernels' configuration are set via the greedy-heuristic and reported in Table 6.5. Therefore, please note that: (1) Some kernels could not be mapped to lower precision in SIMD or MIMD modes with the given accuracy thresholds in the book and such cases are denoted by '-' in the table. (2) If the kernel consists of several sub-kernels, the average utilization of sub-kernels are reported. (3) The utilization ratio is reported for one instance of the kernel. Therefore, one can map more instances of a kernel (in case of available resources) to increase the array utilization and parallelization on the CGRA.

Reconfiguration cycles for the CGRAs: In Table 6.8, the reconfiguration cycles of different CGRA configurations are compared for the implementation of each of the four applications (where K denotes kernel). The number of cycles mainly depends on three factors: 1) Size of the CGRA (some kernels such as Derivative Mask in HCD application might not be mapped entirely on 8 × 8 CGRA and need to be divided to smaller sub-kernels), 2) PE architecture (based on the results of sensitivity analysis and mapping, some PEs can group multiple operations with reduced precision), and 3) application (the number of PEs for required for the execution of some kernel depends on the e.g., image size, cluster nodes, etc.). Considering the K-means classification as an example, the number of PEs depends on the number of cluster nodes. That is the main reason why the reconfiguration overhead has been compared *relatively* for different CGRAs. For the sake of better understanding, the reconfiguration cycles are reported in a parameterized format, i.e., for K-Means the number of nodes is considered to be N and the image size for the image-based applications is considered NxN pixels. For Pan-Tompkins, the ECG sampled are streamed continuously.

Execution Time and Energy Measurements: to calculate the end-to-end execution time of the applications, the widely-used list scheduling algorithm has been adopted for traversing the kernels' DFG nodes with a resource-aware approach [Mic94]. For energy measurements, the energy model and estimation approach of [WCK21] has been followed. In this model, formulated in Eq. 6.1, the energy of computation, memory, and kernel configuration (pre-stored in the memory) are calibrated separately:

$$E_{\text{memory}} = (N_{\text{load}} \times E_{\text{load}}) + (N_{\text{store}} \times E_{\text{store}}) + (N_{\text{not-idle}} \times E_{\text{static}}) + (N_{\text{idle}} \times E_{\text{idle}}) \tag{6.1}$$

Energy of computation: the static power of each CGRA architecture is extracted from the Innovus reports. For the dynamic power measurements, the RTL simulations were performed on a random sequence of data on the post-synthesis net-lists of the CGRA architectures to obtain the CGRA-specific activity files. *Energy of memory and kernel configuration:* for this part, three terms are summed up, as can be seen

Table 6.6 {BB size[1], # PEs} in 8×8 SISD, SIMD, MIMD architectures, for different application kernels

			Harris corner detection	**Pan-tompkins (ECG)**			
Kernel Name/Size	SISD	SIMD	MIMD	**Kernel name/size**	SISD	SIMD	MIMD
RGB to Grayscale (SIMD 4-bit)	{6, 6}	{6, 3}	-	Low-Pass Filter (MIMD 4-bit)	{23, 23}	{23, 12}	{23, 6}
Gaussian Smoothing (MIMD 8-bit)	{63, 63}	{90, 60}	{108, 60}	High-Pass Filter (SIMD 4-bit)	{63, 63}	{63, 31}	-
Derivative Sobel (MIMD 4-bit)	{63, 63}	{90, 60}	{108, 60}	Differentiator (MIMD 8-bit)	{8, 8}	{8, 5}	{8, 5}
Tensor & Score Response (SISD 16-bit)	{63, 63}	-	-	Squarer (SIMD 4-bit)	{1, 1}	{1, 1}	-
Normalization (SISD 16-bit)	{1, 1}	-	-	Moving Avg Filter (SISD 16-bit)	{31, 31}	-	-
K-Means Clustering				**JPEG Compression**			
Kernel Name/Size	SISD	SIMD	MIMD	**Kernel Name/Size**	SISD	SIMD	MIMD
RGB to YCbCr (MIMD 4-bit)	{18, 18}	{18, 9}	{18, 6}	RGB to YCbCr (MIMD 4-bit)	{18, 18}	{18, 9}	{18, 6}
Cluster Cent. Const. (SIMD 4-bit)	{3, 3}	{3, 1}	-	2D-DCT (MIMD 8-bit)	{60, 60}	{75, 55}	{80, 55}
Euclidean Distance Calc. (SIMD 8-bit)	{5, 5}	{5, 3}	-	Quantization (SISD 16-bit)	{64, 64}	-	-
Segmentation/Cluster Cent. Re-Construction (SISD 16-bit)	{14, 14}	-	-				

[1] The number of consecutive instructions, *in the original code*, that are executed per one CGRA configuration

Table 6.7 Array utilization in {SISD, SIMD, MIMD} configuration for different application kernels

	Harris corner detection	Pan-tompkins (ECG)	
Kernel name	**Util. % {SISD, SIMD, MIMD}**	**Kernel name**	**Util. % {SISD, SIMD, MIMD}**
RGB to grayscale	{9.37, 4.68 , -}	Low-Pass Filter	{35.93, 18.75, 9.37}
Gaussian smoothing	{87.89, 64.45, 84.25}	High-Pass Filter	{98.43, 48.43, 23.43}
Derivative sobel	{87.89, 64.45, 84.25}	Differentiator	{12.5, 7.81, 7.81}
Tensor and score response	{72.81, -, -}	Squarer	{1.56, 1.56, 1.56}
Normalization	{1.56, -, -}	Moving Avg filter	{48.43, -, -}
K-Means clustering		**JPEG Compression**	
Kernel name	**Util. % {SISD, SIMD, MIMD}**	**Kernel Name**	**Util. % {SISD, SIMD, MIMD}**
RGB to YCbCr	{28.12, 14.06, 9.37}	RGB to YCbCr	{28.12, 14.06, 9.37}
Cluster Cent. Const.	{4.68, 1.56, -}	2D-DCT	{53.12, 34.37, 28.12}
Euclidean Distance Calc.	{7.81, 4.68, -}		
Segmentation/cluster Cent. Re-Construction	{21.8, -, -}	Quantization	{3.12, -, -}

in Eq. 6.1. These terms are described in the following: 1) *Access energy*: Energy of memory load and store, which are measured separately. 2) *Static (memory-enabled)*: The static energy of memory in one clock cycle is multiplied with the number of clock cycles that the kernel is executing. 3) *Static (memory-disabled)*: The idle energy of memory in one clock cycle is multiplied with the number of clock cycles that it is idle. It should be noted that each of these numbers (for one clock cycle) are looked up directly from the commercial manufacturer datasheets for the 40-nm technology node [WCK21].

Application-Level Comparison of CGRAs: Fig. 6.22 compares the performance metrics of different CGRAs at the application-level, through executing multi-kernel applications. Recalling Table 6.5, the SISD/SIMD/MIMD configuration of PEs for the GREEN CGRA, are adjusted via the proposed greedy-based heuristic. The upper part of each bar (power, energy, or reconfiguration time) in Fig. 6.22 exhibits the share for reconfiguration, while the lower part represents the share of computation. The

Table 6.8 Reconfiguration cycles for CGRA configurations in different application

	Harris corner detection						JPEG compression			
#cycles	K1	K2	K3	K4	K5	Sum	K1	K2	K3	Sum
SISD	$6\times N^2$	$225\times N^2$	$450\times N^2$	$683\times N^2$	N^2	$1365\times N^2$	$18\times N^2$	$960\times N^2/64$	$2\times N^2/64$	$33.03\times N^2$
SIMD	$3\times N^2$	$125\times N^2$	$250\times N^2$	$383\times N^2$	N^2	$762\times N^2$	$9\times N^2$	$512\times N^2/64$	$2\times N^2/64$	$17.03\times N^2$
MIMD	$3\times N^2$	$100\times N^2$	$200\times N^2$	$308\times N^2$	N^2	$612\times N^2$	$6\times N^2$	$512\times N^2/64$	$2\times N^2/64$	$14.03\times N^2$
	K-Means					**Pan-Tompkins**				
#cycles	**K1**	**K2**	**K3**	**K4**	**Sum**	**K1**	**K2**	**K3&K4**	**K5**	**Sum**
SISD	18× N	3× N	5× N	14× N	40× N	23	63	9	31	126
SIMD	9× N	N	3× N	14× N	27× N	12	31	6	25	74
MIMD	6× N	N	3× N	14× N	24× N	6	15	6	22	49

amount for each of the performance metrics is normalized to its baseline counterpart, i.e., of the accurate CGRA. For example, the pair of power values (1.4, 0.2) in the K-Means application in Fig. 6.22 demonstrates that the total power of XbioSip-based CGRA is 1.6× of accurate CGRA, in which the shares of computation and reconfiguration are 1.4× and 0.2×, respectively. Please note that for the sake of visibility, the maximum value in the y-axis of sub-figures are set to 1, as only the XbioSip-based CGRA has lower performance than the accurate CGRA. The following deductions are notable based on the results.

First, although the architecture-level gains of GP-CGRA and X-CGRA (from Table 5.4) are also reflected at the application-level, ultimately the SIMD/MIMD modes of GREEN enable the highest performance gains by implementing multiple functions of reduced-precision in a smaller number of PEs. This means that GREEN offers substantial application-level savings, while maintaining an acceptable QoR.

Second, as can be seen in the figures, the overhead of reconfiguration time and energy not only depends on the application structure but also on the CGRA architecture. In fact, for each of the applications, the SIMD/MIMD version of GREEN CGRA requires lower reconfiguration time and energy, as kernels are precision-reduced and less (groups of) data needs to be communicated to/from the memory. For example, such overheads are the lowest for the Pan-Tompkins application, in which the average precision of kernels are more reduced (kernel structure of [4-4-8-4-16]) and highest for the JPEG Compression, in which kernels are less precision-reduced ([4-8-16]).

Third, GREEN CGRA can be beneficial for edge computing, as processing at a lower precision not only contributes to the savings in the memory energy (both for reconfiguration and data access), but also, the end-to-end latency of the application can be reduced. This latency reduction is due to the shortened propagation delay of precision-reduced individual operations (in application cases that all PEs are configured to operate at 4- or 8-bit precision). As the result, the clock period of the application/CGRA and thereby, the execution time of the application will be reduced. In fact, improving the application speed and energy is highly desirable for enabling real-time processing at the edge (e.g., heart anomaly detection in wearable 24/7 gadgets). Enabling such a shorter response time and reduced energy consumption through processing and storing data in the proximity of data is referred to as Multi-Access Computing Error (MEC) [Abd+19], which can provide many advantages for battery-operated devices. From energy and real-time processing perspectives, MEC enables pre-processing and filtering or extraction of necessary features at the edge (rather than transferring the whole sampled data to the cloud). From data privacy perspective, private information of users can be preserved at the edge, e.g., in smart-health gadgets. It is worth highlighting that the unutilized PEs of GREEN CGRA (when running the program in SIMD/MIMD modes) can be power-gated, or leveraged for analysis of data at a higher sampling rate or accommodation of more tasks.

6.5 Conclusion

In this chapter, application-level and cross-layer approximation methodology schemes were presented and evaluated on both FPGA and CGRA reconfigurable fabrics. In this context, an end-to-end error-resiliency sensitivity analysis was also presented, serving as a prerequisite for the efficient implementation of the cross-layer approximation methodology.

This analysis has highlighted four observations as key guidelines for our methodology: 1) It is unnecessary to dedicate the high-precision of 32-bit uniformly to all kernels. 2) Kernels significance order in lost QoR can differ from its order in the gained performance when approximated by the same techniques. 3) Multi-kernel applications usually show higher sensitivity to approximation of addition operation than multiplication. 4) Low EB has a high impact on the final QoR, as errors with different signs can nullify each other, provided that kernels have aggregation-based structure (i.e., having mostly addition and multiplication operations).

Based on these observations, a cross-layer approximation methodology was proposed, employing precision scaling (application-level technique) as well as inexact configurations for the multiplier and divider (circuit-level technique). For adjusting the approximation knobs, a greedy-based heuristic was proposed. This heuristic adopts the $\frac{\Delta\text{Performance-Gain}}{\Delta\text{QOR}}$ as the saliency/deciding metric that appropriately reflects intensity of gained performance over a possible accuracy loss when applying a chain of approximations. This DSE technique has been able to efficiently tune the approximation knobs and swiftly generate Pareto- or near-Pareto configurations which render optimal- or near-optimal trade-off between performance gain and accuracy loss. The proposed heuristic also obviates the need for an exhaustive search in DSE cases having multiple knobs.

Our methodology has been evaluated on multiple single- and multi-kernel applications and demonstrated significant savings. For example, in an FPGA-based system, the end-to-end evaluations of non-pipelined RAPID, deployed in three multi-kernel applications in the domains of bio-signal processing, image processing, and moving object tracking for UAV indicate up to 35%, 33%, and 45% improvements in area, latency, and ADP, respectively, over accurate kernels, with negligible loss in QoR. Moreover, the end-to-end evaluations of SIMDive and the cross-layer sensitivity analysis and greedy heuristic on the same applications have shown up to 64%, 62%, and 86% improvements in area, latency, and ADP, respectively, over 32-bit fixed precision, with marginal accuracy drop. Finally, the experiments conducted on GREEN CGRA not only corroborate the similar performance gains, especially for the MIMD mode (e.g., speeding up the heart QRS detection by 49% and consuming 43% less area and 66% less energy than an accurate CGRA) but also maintains the heartbeat detection accuracy at 100%.

Chapter 7
Toward Cross-Layer Approximation for the Distributed and In-Network Acceleration of Multi-kernel Applications

Abstract Enabling remote computation for continuous signal processing workloads—from bio-signals to image and audio—requires both high processing speed and secure, low-latency data handling. Relying solely on cloud computing increases network traffic and cannot ensure robust privacy protection [Abd+19]. These constraints have motivated the shift toward INC, where data is processed while traversing the network [Sap+17]. Although programmable network devices now offer enhanced flexibility for such computation, their limited memory, restricted arithmetic support, and lack of FP capabilities hinder multi-kernel deployments. This chapter introduces X-DINC, a cross-layer approximation and distribution methodology that exploits application-level error resilience to overcome these limitations. Building on Plasticine (Chap. 6), X-DINC identifies optimization opportunities in data acquisition and computation, simplifies complex operations for INC, analyzes sensitivity between performance and QoR, and applies a greedy heuristic to generate near-Pareto configurations that maximize performance while preserving user-defined accuracy.

Keywords Approximate computing · Cross-layer approximation · Reconfigurable computing · In-network computing · Distributed computing · 5G-6G · Fpga · High-throughput · Edge computing · Sustainability · Energy-efficiency · Area optimization · Mitchell's algorithm · Logarithmic hybrid multiplier-divider · Error sensitivity analysis · Multi-objective heuristic · Design space exploration · Blind source separation · Acoustic anomaly detection · Independent component analysis · Programmable networks · P4

7.1 Introduction

Enabling remote computation for many 24/7 signal processing applications, from bio-signals to image and audio, requires high processing speed as well as secure and swift data transportation. Relying solely on conventional cloud computing not only imposes heavy traffic on the network but also fails to guarantee patient privacy protection [Abd+19]. These challenges have driven the need for processing data

Z. Ebrahimi and A. Kumar, *Cross-Layer Approximation and In-Network Acceleration*,
https://doi.org/10.1007/978-3-032-21712-7_7

on the fly while it is being transmitted within the network, a paradigm known as INC [Sap+17].

To address this need and keep pace with rapid advancements in computational approaches across diverse applications, it is crucial to leverage the evolution of programmable network devices. These advancements enable network infrastructures to transition into computing pipelines, providing INC capabilities through improved hardware and enhanced programmability. By processing data in-network rather than relying solely on centralized cloud computing, INC reduces energy consumption and network congestion, contributing to more energy-efficient and sustainable computing.

However, despite the initial success in offloading single/small kernels to the network devices, utilizing INC is still in its infancy and faces main challenges for multi-kernel applications. In particular, the limited memory and computing resources, along with the lack of support for FP and complex operations, have hitherto prohibited the deployment of multi-kernel applications on resource-constrained network devices.

To tackle the aforementioned challenges, a cross-layer approximation and distribution methodology (X-DINC) is presented, exploiting the error resilience of applications. By expanding the Plasticine methodology (presented in Chap. 6), X-DINC effectively utilizes the synergistic effects of a chain of techniques to facilitate the deployment and distribution of kernels over heterogeneous devices in an INC environment. To this end, various approximation and optimization opportunities are first identified in both the data acquisition and computation phases of multi-kernel applications. Second, the structure of a set of complex arithmetic operations is simplified to cope with the *computation* limitations of programmable network switches and tailored for an INC implementation. Third, an application-level sensitivity analysis is applied to measure the correlation between end-to-end performance gain and the loss of QoR, after individual kernels are approximated by various techniques. Finally, a greedy heuristic algorithm is employed to swiftly generate Pareto /near-Pareto mixed-precision configurations that maximize the performance gain while maintaining the user-defined QoR.

X-DINC is prototyped on an Xilinx Virtex-7 FPGA and evaluated on the widely used BSS application on the real-world industrial audio dataset. The evaluation results confirm that compared to an *Accurate-Centralized* approach, X-DINC performs the separation job up to 35% faster with up to 88% less ADP, respectively, when the task is distributed on 2–7 network nodes. These gains are achieved when the audio quality is maintained within the acceptable range of 15–20 dB.

To the best of our knowledge, this is the first work that attempts to enable the in-network acceleration of multi-kernel applications with the goal of minimizing response time (while guaranteeing an acceptable QoR). In the following, the motivations for adopting a cross-layer approximation strategy to enable the execution of multi-kernel applications in an INC environment are elaborated.

7.2 Cross-Layer Approximation Opportunities in Distributed and INC

Figure 7.1 illustrates the cumulative energy of the addition operations, compared to multiplication and division in the case study BSS application. As can be seen in this figure, the share of energy for addition operation is negligible compared to the cumulative energy for the aforementioned complex operations. Moreover, recalling Fig. 2.5 of Sect. 2.3, not only the area, latency, and energy of accurate multiplication, division, and square root operations are significantly larger than those of addition of the same size but also these performance metrics grow exponentially when the operand size is increased. Furthermore, neither such resource-hungry operations, nor the FP data type are supported in the programmable ASIC switches in the 5G networks. Finally, as discussed in Sect. 1.2, the real-world sampled data for such streaming applications are often noisy and/or carry redundant information. Therefore, skipping parts of the samples many not affect the final QoR of the application. These insights motivate a cross-layer approximation scheme that optimizes data acquisition and computation phases in multi-kernel application execution. In the following, the adopted techniques are discussed and the sensitivity of a widely used 5G application case study to such techniques is analyzed.

7.3 X-DINC Error-Resiliency Sensitivity Analysis

As discussed in Sections 3.9.3 and 3.9.4, cramming multi-kernel applications into network devices is neither straightforward nor trivial. In fact, an accurate and high-precision implementation may not even be possible due to both resource constraints and the limitation of the P4 language. Therefore, to address such issues, error-resiliency opportunities across the layers of abstraction were investigated and capitalized on. To this end, as also shown in Fig. 7.2, several approximation/optimization knobs are identified in Sect. 7.3.1, and in Sect. 7.3.2 a generic methodology is pre-

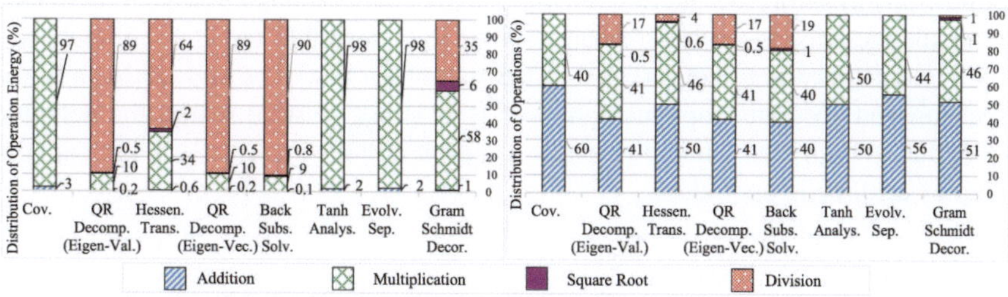

Fig. 7.1 Cumulative percentage and energy of each operation in the kernels of BSS application, 32-bit fixed-point implementation

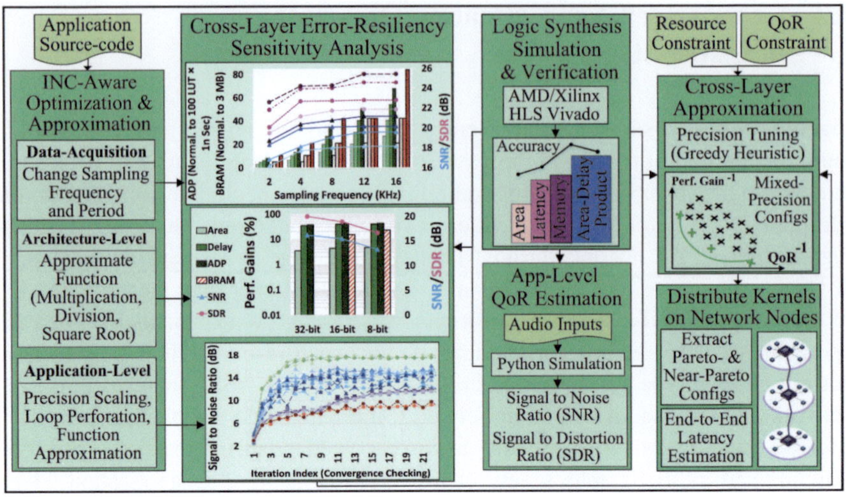

Fig. 7.2 X-DINC framework: Overall design, approximation, and distribution methodology for an INC environment

sented to measure the error sensitivity of application kernels to various approximation techniques.

The sensitivity analysis results are then imported to the proposed approximation heuristic of X-DINC (presented in Sect. 7.4), where the goal of the algorithm is to maximize the performance gain (i.e., improvements in resource footprint, energy, or response time of application) while maintaining an acceptable QoR via operating the possible knobs. Finally, as the ultimate goal, the approximated configurations of kernels are distributed over a chain of network elements in a way that minimizes transmission latency (thereby also minimizing the application response time). This is particularly important because the existing INC approaches (cited in Section 3.9) may be impractical for a single-node deployment due to limited resource capacity or in a distributed setting due to the transmission overheads.

In this context, BSS problem is a symbolic case study among other multi-kernel applications to showcase the efficacy of X-DINC methodology since it inherits all the challenges discussed before and is cited as a desirable target for in-network acceleration (recalling Section 3.9). In the following, it will be shown that a high-precision and accurate implementation of BSS is not feasible within the network nodes due to both their limited resource footprint and the inability of P4 to support complex arithmetic operations.

Table 7.1 The leveraged cross-layer approximation knobs in the sensing and computing phases of applications, for a distributed and INC-setting

Data acquisition (sensing)	Computation	
Sampling period	Architecture-level	Application-level
Sampling frequency	Approx. Multiplication	Precision scaling
	Approx. Division	Loop perforation
	Approx. Square root	Approx. function (tanh)

7.3.1 Cross-Layer Approximation Knobs

As listed in Table 7.1, various approximation and optimization opportunities were identified in the data acquisition and computation phases of the applications executed in an INC environment. In this book, these knobs—adopted from different layers of abstraction in the computing stack—target the optimization of both the pre-processing and processing kernels of the BSS application[1].

Sampling period and frequency: Analog signals around us, such as audio and bio-signals, are sampled and converted to the digital domain using an ADC. These signals often contain substantial correlated or redundant data, partly due to their periodic structures. Thus, reducing the number of samples can significantly improve these applications' processing requirements and response time [RBD13]. In previous studies [CS05, He+22, Wu+23, Wu+21d], the adopted sampling period and frequency for the implementation of audio-based BSS range from $\sim$ 1-10 seconds and 4-16 KHz, respectively. Therefore, as the first knob, the consequences on the QoR and performance metrics of changing the sampling frequency (2–16 kHz) and period (4–10 seconds) are proposed to be analyzed. The result of this trade-off is presented in Sect. 7.5.2.

Architecture-level: For the second set of knobs, some basic but ubiquitously used arithmetic operations (i.e., multiplication, division, and square root) are approximated based on techniques with low EB. Choosing these operations targets to overcome the inability of P4 to implement them in ASIC switches; in addition, these operations require significantly higher area, delay, and energy compared to simpler ones such as addition and multiplication (the comparison results can be found in Figure 2.5). Note that the addition and subtraction operations were not approximated because: 1) The accurate implementation of addition/subtraction is already supported in ASIC switches, 2) the cumulative energy of the operations is negligible (e.g., in BSS application), compared to the cumulative energy for aforementioned complex operations (see Fig. 7.1), and 3) both previous studies [HJi+20, Ebr+21] and our sensitivity analysis showed that the approximation of addition in matrix-multiplication based kernels may result in high accuracy fluctuations, especially when the errors of the inexact adder are biased toward the same sign. Besides, these

[1] These knobs are not exclusive to the BSS; they exist and can be leveraged in a wide range of applications.

situations also hold true for other classification, bio-signal, and image-processing applications [EK24, Ebr+21, EK21, Ebr+22].

Application-level: Three techniques are considered at this level. First, *precision scaling* is adopted, as it is a widely used approximation technique. It has been tested on a broad range of programs [SSB18, Luo+22, Lui+21, XZ19, SB18], where significant resources can be saved in both computing and memory aspects. In addition, processing at a lower precision can also reduce the application's end-to-end latency because of the shortened propagation delay of the individual (precision-reduced) operations. Moreover, *loop perforation*, i.e., skipping loop iterations, is another effective method that was applied, for example, to the BSS application (in the Gram–Schmidt decor relation, Back Substitution solver, and tan Hyperbolic – tanh – analysis kernels, as shown in Fig. 2.11). Last but not least, *function approximations* are applied to simplify the implementation of complex functions, thus further increasing the resource efficiency. For example, to approximate the tanh function, the highly accurate implementation of Python `PiecewiseLinFit` from `pwlf` library was approximated via a piecewise linear segmentation approach using 4, 8, and 16 segments. Finally, the 8 segments have been adopted because of their negligible accuracy difference compared to the IP-based implementation of tanh function in Xilinx Vivado [Xil24]. Another example is the Gram-Schmidt de-correlation, which is implemented based on Scikit-learn open-source python library [Laf24] and further modified for a resource-efficient implementation.

Next, the three types of optimization knobs are tested individually and jointly to understand the (synthetic) effects on the performance of the BSS application.

7.3.2 End-To-End Kernel-Wise Sensitivity Analysis

This analysis is motivated by the observations in previous studies [Tas+20, Ebr+21, EK24] which state that not only kernels contribute differently to *each of the* performance metrics (e.g., application latency, area, or memory footprint) but also after approximation, their significance to the performance gains might differ from their importance influencing the error-resiliency.

The sensitivity analysis performed in X-DINC is inspired by [Ebr+21] that presented in Sect. 6.2. However, in [Ebr+21], only the approximation of multiplication and division operations and precision scaling were considered. This chapter further expands the scope in the following aspects. First, analyzing the sampling frequency and period needs to be considered in a distributed INC-setup. This is because the data is usually gathered from multiple sources (e.g., sensors), and changing data size can significantly affect the time and/or resources required for processing. Second, our analysis assessed the memory footprint, which is another constraint in various network nodes, but missed out in [Ebr+21]. Third, additional approximation techniques were evaluated in our analysis. For example, loop perforation will significantly affect the response time of an application, and simplifying the square root operation and more complex functions is also necessary, as they are not supported in ASIC switches. Fourth, the error metrics considered in this book are expanded to SNR and SDR (originally was PSNR in [Ebr+21]). In fact, SNR and SDR not only consider the background noise but also reflect the effects of distortion and unwanted artifacts due to both compression and various approximations [Wu+21c]. Finally, the *integer to integer* precision scaling approach of [Ebr+21] is changed to a *FP to fixed-point* scaling, to also consider the effect of *data type approximation*.

Our sensitivity analysis contains four steps, described below (the results are detailed in Sect. 7.5.2). ① The performance–accuracy trade-off for executing the whole multi-kernel BSS application at different sampling periods and frequencies is reported, showing that processing at a high data rate requires a significant amount of resources (numerical comparison can be observed in Fig. 7.4). This hinders or exacerbates the difficulty of executing multi-kernel applications on resource-constrained network nodes. ② The accurate arithmetic multiplication, division, and square root operations and tanh functions are replaced with INC-tailored (inexact) alternatives in all application kernels, and their marginal effect on the final accuracy is demonstrated (see Fig. 7.3). ③ To apply mixed-precision tuning on consecutive kernels, a *kernel-wise* precision scaling investigation was performed (as multi-kernel applications usually exhibit a more diverse error-resiliency spectrum to the precision scaling technique [Ebr+21, Tas+20, EK24]). To this end, the precision scaling is applied—on top of the previous techniques—on each kernel individually, while the rest of the kernels are accurate (32-bit fixed point precision with *exact* operations). The result of precision scaling analysis is also depicted in Fig. 7.3. ④ Finally, the effect of loop perforation (loops used to calculate and update the separation matrix in an iterative approach) on the accuracy of the application is analyzed. Specifically, the effect of reducing the loop iteration count from 200 in a highly accurate implementation [Wu+21c] to 1 is examined. The loop perforation is applied on top of the previous techniques, and the result shows that the application successfully converges after 35 iterations, even when it is implemented with reduced precision and with lower sampling frequency and period (see Fig. 7.5). The output of this analysis is multiple lists, each belonging to a specific kernel, and reports a pair of $\{\Delta\text{ADP}, \Delta\text{QoR}\}$ that reveals the end-to-end performance-accuracy trade-off when only that kernel is approximated while the rest are accurate.

7.4 X-DINC Cross-Layer Approximation and Distribution Methodology

7.4.1 Greedy-Based Approximation Heuristic

Targeting a high accuracy, computations can be carried out in FP precision. However, as discussed earlier, reducing the precision to, e.g., 16-bit has also been reported to be satisfactory for many applications, including BSS [Odo12].

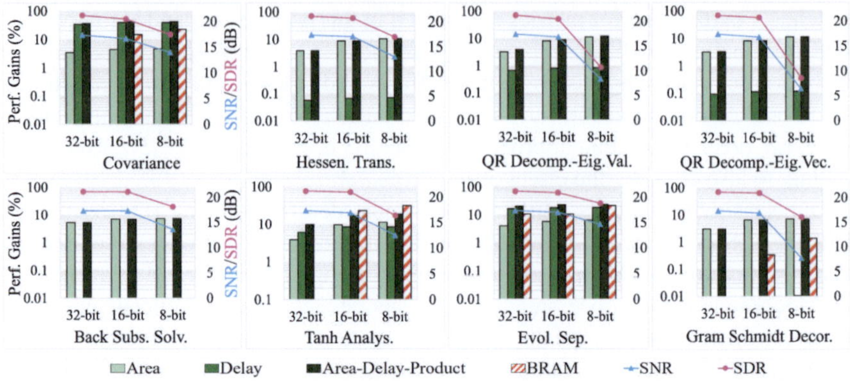

Fig. 7.3 Sensitivity analysis of BSS application to cross-layer approximation techniques, i.e., precision scaling on top of function approximation (multiplication, division, square root, tanh). The figures show the trade-off between the end-to-end Δperformance and ΔQoR after approximating each kernel, *individually* (while the rest are accurate FP precision)

The main goal of the cross-layer approximation methodology is to appropriately adjust the precision of each kernel (on top of other techniques applied to the kernels) in order to maximize the performance gain while minimizing the end-to-end quality loss (or maintaining a predefined accuracy threshold). The importance of such methodology becomes pronounced for preventing the need for an exhaustive DSE for multi-kernel applications, especially at run-time. In this regard, to tune the precision of consecutive kernels in a multi-kernel application, the greedy-based heuristic of [Ebr+21] was customized and further improved. Note, the greedy heuristics have lower complexity than other conventional choices such as ILP and GA and faster convergence to the near-optimal solutions than simulated annealing approaches [Ebr+21]. These are highly important factors for runtime decision-making actions in INC scenarios.

The inputs of the heuristic are the predefined QoR threshold and the information obtained from the sensitivity analysis, i.e., two lists L_1 and L_2 that detail the end-to-end performance gain and QoR-loss for down-scaling the precision of the kernel to either 16- or 8-bit (while other approximations are already applied on that kernel). Each item of these lists reveals the end-to-end gain/loss only for one approximate kernel, while the rest of the kernels are 32-bit precision. The customization of the heuristic applied in this book involves reducing the number of required lists from one list of {*performance-gain*, *QoR loss*} (for each approximation technique [Ebr+21]), to only two (which shows the performance-gain and accuracy trade-off for 16- or 8-bit kernels), overall. This means that in the baseline configuration (input of the heuristic), the loop perforation, operation/function approximations, and reducing sampling period/frequency are already applied. Therefore, the heuristic itself only determines the precision of the kernels. The reason is based on the sensitivity analysis result: the precision scaling technique has shown a more pronounced effect on the kernels than other techniques (see Figs. 7.4 to 7.5).

Modification to the original heuristic (i.e., Algorithm 1), *presented in Sect.* 6.3: In this section, the techniques were combined (those having a lower impact on the ultimate application accuracy). This has significantly reduced the size of the design space and, hence, the exploration time of the heuristic. Pruning the dominated points in the final design space that needs to be explored is highly desirable, especially for runtime decision-making in the 5G and upcoming 6G networks. In fact, in runtime scenarios, heterogeneous nodes are dynamically added or removed from the network, necessitating a quick redistribution of application kernels to the updated network nodes.

The rest of the heuristic is similar to the original one and runs as follows: First, all application kernels (already approximated by other techniques except the precision scaling) are uniformly set to 32-bit precision. Afterward, the *Salience List* is constructed by adding the $\frac{\Delta \mathrm{ADP}}{\Delta \mathrm{QoR}}$ for each pair of {kernel, precision} (i.e., the precision of that kernel is reduced to 16- or 8-bit). Note that this value is obtained in the sensitivity analysis phase. The list is then sorted in a descending order to reveal the significance order of kernels for the *combined* approximations: The kernels at the top of the list can enable a higher end-to-end performance gain while imposing less QoR loss to the ultimate application accuracy (when approximated by all the techniques). Based on this list, the greedy heuristic is applied in an iterative manner: In each iteration, the pair of {kernel, precision}, which appears at the top of the salience list, is chosen for precision scaling. After generating a new configuration of kernels at each step of the heuristic, the configuration is evaluated on diverse samples to assess whether the final desired QoR threshold is preserved. Whenever the accuracy of the generated configuration crosses the threshold (having up to 5% difference with the user-defined threshold), the heuristic backtracks to the previous accuracy-satisfied configuration and continues the search by evaluating the next candidate at the top of the salience list.

7.4.2 *Distributing Approximate Kernels over Heterogeneous Network Nodes*

A distributed approach in an INC setup entails decomposing the application into multiple partitions, each of which is assigned to a network device[2], and the network devices are connected with a network topology [Zhe+23a, Sul+21]. For such a distributed setting, X-DINC methodology needs to distribute the N kernels across M network nodes ($M \in \{2, ..., N-1\}$) by finding an optimal or near-optimal grouping. The goal is to minimize the transmission latency by identifying the best partitioning points that minimize the size of the transmitted data (such a goal has neither been considered nor explored in any of the related works cited in Sects. 3.9.4 and 3.9.5).

[2] Finding the optimal re-partitioning and re-distributing application kernels w.r.t. network dynamic changes (e.g., in distributed training or inference of NNs) is an interesting follow up track for this book.

In this context, our approximation approach helps find an optimal or near-optimal distribution solution: It is proposed to focus only on those Pareto or near-Pareto configurations (generated by the heuristic) that not only render the highest ADP gain for a user-given QoR threshold but also contain more kernels with the lowest precision (i.e., 8-bit for this book).

Recall that Algorithm 1 tunes the precision of kernels in a way to minimize the $\{resource \times latency\}$ of the application (while satisfying the user-given QoR-level). Therefore, multiple Pareto or near-Pareto configurations are already generated, with minimum or near-minimum latency for a given accuracy threshold (see Fig. 7.6). It is worth highlighting the unique features enabled by the X-DINC that are not supported by the existing works: Through reducing the size (bit-width) of intermediate data that need to be communicated; our methodology minimizes the transmission latency (which could be the dominant portion in the end-to-end latency of the application, as discussed in Sect. 1.6). Moreover, X-DINC enables the flexibility to choose among *multiple* kernel configurations having an admissible accuracy. In fact, each of these (mixed-precision) configurations might be suited for a specific distribution scenario in an INC-setup, in which network nodes have heterogeneous resource footprints. To show the efficacy of the proposed distribution approach over the centralized counterpart, various scenarios with different numbers of network nodes and QoR thresholds are assessed in Sect. 7.5.4.

7.5 Results and Discussion

7.5.1 Experimental Setup

Application partitioning and mapping We developed the C++ implementation of BSS application based on [Wu+21c, PUL22] and afterwards, it was synthesized with Vitis HLS 2020 on a commodity FPGA (Xilinx Virtex-7 VC709). For the exact performance analysis, the HDL-generated design from HLS was further passed to the downstream implementation phase, placed and routed on Virtex-7 through Xilinx Vivado. In the experiments, the DSP units are disabled, and only the 6-LUTs are used for the synthesis. The reason behind this is two-fold. First, utilizing only 6-LUTs eases the comparison of different approaches in terms of area (as estimating the required number of LUTs to be replaced with a DSP varies for each function and hence, is not a straightforward task). Second, adopting an LUT-based implementation is also recommended by many FPGA vendors for low bit-widths operations (e.g., at 8-bit) [AMD21]. It should be mentioned that the kernels include all the computational segments of the application, and only the non-computational and critical parts (e.g., memory or loop index calculation) are exempted from the approximations. Finally, as shown in Figure 2.11, the application is partitioned into eight computational kernels and analyzed by the sensitivity analysis process.

Benchmark Similar to state-of-the-art works [Wu+23, Wu+22c, Wu+21d, Wu+21c], the widely used acoustic dataset of MIMII [Pur+19], which collects real-world data from industrial machines, was utilized. MIMII contains 26092 normal and abnormal operating acoustic data from four types of machines: Valves, pumps, fans, and slide rails. Every segment has a duration of 10-second audio, sampled at the frequency of 16 KHz.

QoR metrics The accuracy of an approximation configuration is assessed with different metrics and simulations. SDR and SNR were measured as the indicators of separation quality for the audio-based BSS application. These metrics are the most widely used audio accuracy metrics in the BSS processing studies, comprehensively covering different types of errors [Wu+21c]. SDR and SNR are already defined in Eqs. 2.13 and 2.14 respectively, wherein S and $\hat{S}$ denote the original and estimated source. e_{interf}, e_{noise}, and e_{artif} represent the errors attributed from interference, noise, and artifacts, respectively.

7.5.2 *Cross-Layer Error-Sensitivity Analysis Results*

In the left-hand side of Fig. 7.4, the contribution of each kernel is first detailed when the BSS application is implemented in accurate mode. The evaluated metrics are the end-to-end latency (the summation of kernels' delays), 6-LUT count, and ADP as a representative of energy[3]. This figure shows that the kernels contribute differently to each of the performance metrics (e.g., application latency or energy). It is important for the designer to consider this information when targeting a high-performance or energy-efficient design.

In the following, the results of the sensitivity analysis are presented. As discussed in 7.3.2, this process is applied in four steps. The right side of Fig. 7.4 demonstrates the result of the first step, i.e., the sensitivity (i.e., separation accuracy) of BSS to various sampling periods and frequencies. Afterward, Fig. 7.3 reveals the low sensitivity of individual kernels to replace the complex operations/functions in all kernels with the approximated alternatives. Moreover, this figure also shows the sensitivity of individual kernels to precision scaling by applying this technique with a one-kernel-at-a-time strategy discussed earlier. Finally, Fig. 7.5 shows the effect of loop perforation (on top of other approximation techniques) on the accuracy of the application and proves the convergence of the separation matrix for all the scenarios. The key observations from the sensitivity analysis are pin-pointed herein.

Changing the sampling period and frequency: As can be observed in the right-side of Fig. 7.4, implementing the BSS application at the high accuracy of FP precision with high sampling frequency and period (similar to [Wu+21c, Wu+23, Wu+22c, Wu+21d]) is significantly resource-hungry. For example, the right-most Green bar and Red bar in this sub-figure represent the requirement of 50860 LUTs and 84

---

[3] A comprehensive and detailed energy calculation for such a large multi-kernel application is arduous and considered for the follow-up track.

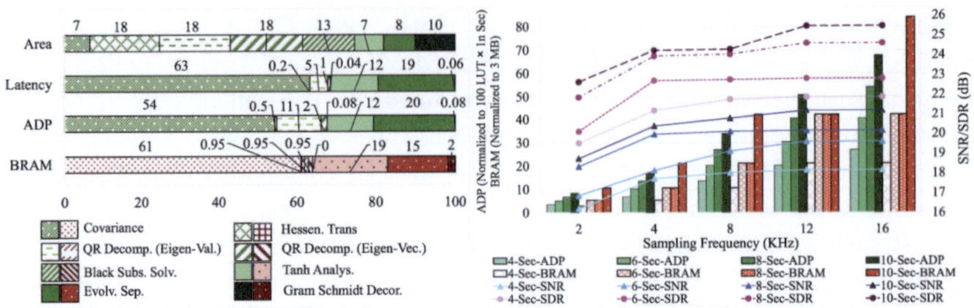

Fig. 7.4 The contribution of each kernel in the total performance metrics of the BSS application (left). The effect of changing sampling period and frequency on the accuracy and performance metrics of BSS application (right)

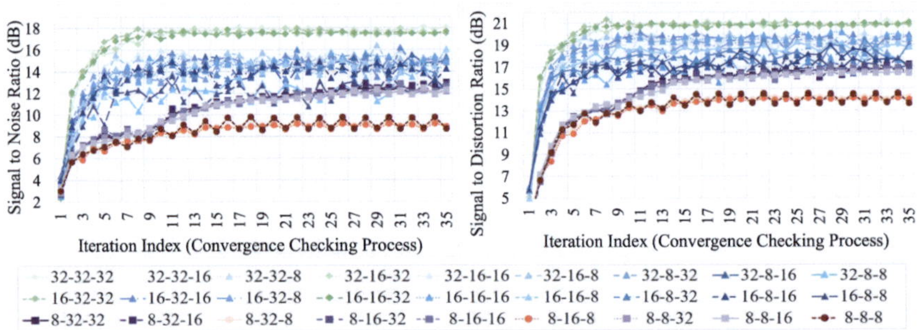

Fig. 7.5 The effect of loop perforation technique on the convergence and accuracy of the BSS application

MB memory, respectively, for implementing the BSS configuration that can support processing a 10-second audio sampled at the frequency of 16 KHz. Such a resource-hungry implementation hinders the practical deployment of this multi-kernel application in most of the FPGA- or ASIC-based network switches. In fact, only a few FPGAs from Xilinx Versal, UltraScale, or Virtex have a memory capacity of higher than 30 MB. However, as can be observed in this figure, reducing the sampling frequency and period to 4 s and 4 KHz (Green and Red bars on the left side) not only reduces the required resources significantly (up to 93% reduction in memory and 42% reduction in 6-LUTs), but also maintains the accuracy at an acceptable level (> 16 dB) and satisfies the Nyquist-rate sampling criteria [Nyq28]. It is worth noting that the SDR/SNR value above 15 dB is considered as an acceptable separation quality for BSS application [ZZ13, Wu+20b, KK14]. Therefore, based on this observation, an audio period of 4 s, sampled at a frequency of 4 kHz, was chosen for the rest of the experiments.

*Approximation of operations (multiplication, division, square root) and functions (*tanh*)*: As can be seen in Fig. 7.3, replacing the exact operations with the SIMDive-based version [EUK20] (which is P4-compatible[4]) not only can significantly improve the ADP of the kernel but also marginally affects the accuracy in most kernels, when the precision is 32- or 16-bit (all bars with 32- or 16-bit label have a high SDR/SNR accuracy of more than 16 dB). In fact, even when these operations/functions are approximated in all application kernels, the SDR (SNR) metrics are higher than 18 (14) dB (see the black hexagon markers in Fig. 7.6). Our analysis has shown that this negligible quality drop is endowed with the near-zero biased errors of SIMDive multiplication/division, which can cancel out each other in consecutive kernels. Moreover, as mentioned earlier, cutting-edge studies have also shown that a low error-bias error plays a pivotal role in minimizing the error accumulation in applications having multiple kernels with an aggregation-based structure (i.e., mostly addition/multiplication operations) [Zer+21, EK24, Ebr+21, Ebr+22, ZSA20]. Based on this observation, SIMDive-based operations have been deployed in all pre-processing and processing kernels of the BSS application. The replacement of accurate multiplication/division with SIMDive, in all kernels, has pruned the dominated points and efficiently reduced the size of design space that needs to be explored, in which only the precision of kernels is required to be changed (see Sect. 7.5.3).

In contrast to the approximation of multiplication and division, our analysis shows that truncating the output of the addition/subtraction operations (more than a few bits) can significantly affect the accuracy, especially when applied in kernels such as Covariance. The high sensitivity of addition to approximation for other matrix multiplication-based applications is also endorsed by [HJi+20]. Therefore, addition and subtraction operations are maintained accurately in our implementations. Moreover, the small contribution of these operations in the total energy of application kernels (Fig. 7.4) justifies our decision that it is better to preserve the accurate structure of these operations.

Kernels contribute differently to the QoR fluctuations and the gained performance: as discussed earlier, the kernels contribute differently to each of the performance metrics (Fig. 7.4 left side). Therefore, the importance order of kernels (for approximation) would be different, w.r.t. the ultimate design goal. For example, Fig. 7.3 shows that when a high-performance design with a shorter separation time is the goal, approximation of the *covariance* kernel results in higher latency reduction (due to its division, acting as the speed bottleneck operation). However, when resource (LUT) saving is the goal, the approximation of *Hessenberg* and *QR Decomposition* kernels is more beneficial. Moreover, this figure also reveals that the significance order of kernels in their *QoR loss* can vary from their order in the *gained performance* (after applying the same approximation technique). For example, as can be seen in Fig. 7.3, the most error-resilient kernel is *Evolving Separation*, while, in general, the approximation of *Covariance* can result in higher performance gain.

Overall, aforementioned insights further corroborate our initial statement, i.e., that the *relation* between ΔPerformance and ΔQoR should be considered for adjusting

[4] Further reasons for using SIMDive MUL/DIV is discussed in Appendix.

the approximation knobs in a multi-kernel application and not based on the primary appearance order of kernels (or their contributions), in the accurate configuration.

Effect of loop perforation on accuracy and convergence of separation matrix: Fig. 7.5 shows that the accuracy fluctuations and convergence of the separation matrix after applying the loop perforation (on top of other techniques) on the last three kernels of BSS (which consist of loops and can be tuned to 8−, 16−, or 32−bit precision). As can be seen in this figure, reducing the number of iterations in the convergence checking to $\sim$ 25-35 not only maintains an acceptable QoR-level for many of the kernel configurations (i.e., the configurations that maintain the SDR above 15 dB), but the separation matrix also successfully converges. Skipping further and unnecessary iterations can greatly reduce the application's response time. Therefore, the separation matrix was updated up to 35 times in the experiments.

The outputs of the sensitivity analysis (the lists of end-to-end $\{\Delta$performance gain, ΔQoR$\}$) are given to the precision-tuning heuristic in the next step.

7.5.3 Performance Gains of Approximate Configurations Obtained by the Cross-Layer Approximation Methodology

Herein, the overall efficacy of the cross-layer approximation methodology is evaluated. Figure 7.6 illustrates the accuracy and performance gain of approximate configurations over the baseline double-precision FP configuration having the accurate operations (each of the eight kernels can have the precision of 8-, 16-, or 32-bit, resulting in $3^8 = 6561$ approximate configurations). Moreover, Fig. 7.6 also demonstrates the efficiency of the greedy heuristic in finding the Pareto- or near-Pareto points by distinguishing the ones that are found by the heuristic in different exploration time limits. The results show that not only new ADP-QoR trade-off levels (mixed-precision configurations) are generated by the X-DINC methodology, but also these approximate configurations reach a higher performance gain than the uniform-precision ones. Please note that not only one unique but a set of configurations reach permissible ADP-QoR trade-offs, which can be utilized for network nodes having different resource footprints. For example, it is worth highlighting that the proposed X-DINC methodology has been able to generate the Pareto-optimal configuration of the BSS application which can be fit into one single FPGA (in terms of both memory and LUT) while maintaining an acceptable accuracy of $>$ 16 dB (see overlapping pink circles and green stars in, e.g., sub-figures at the right side of Fig. 7.6). can also support dynamic scenarios in multi-hop networks and energy management routines for heterogeneous resources: Arbitrary heuristic-generated configurations can be stored in FPGA SPI/BPI flash memory and loaded at runtime, enabling X-DINC to adapt to, e.g., network congestion, node failures, or bandwidth changes, by switching configurations and redistributing approximated kernels.

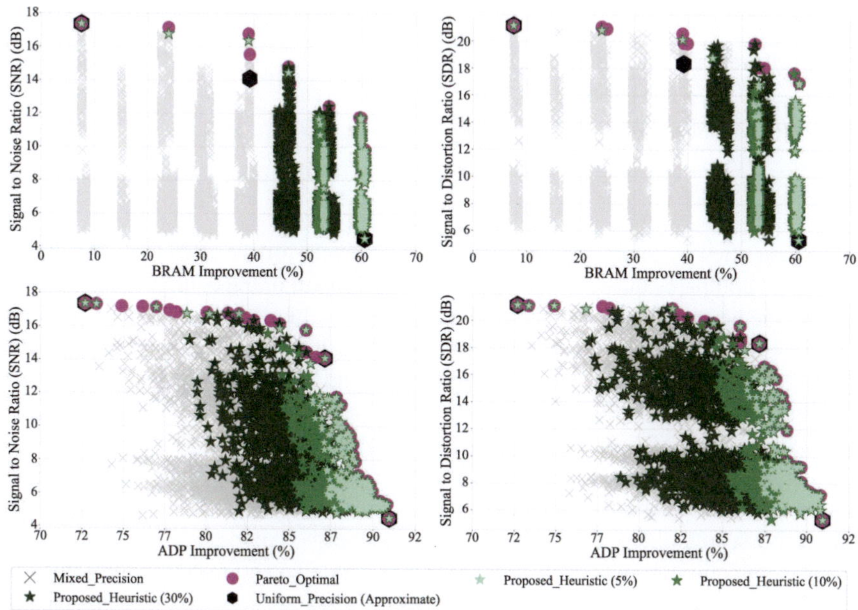

Fig. 7.6 Different performance-QoR trade-offs are enabled by the *approximate, mixed-precision* configurations, which also achieve similar/higher gains compared to *uniform* configurations. The heuristic also finds many Pareto/near-Pareto points in the 10%, 20%, and 30% time of the exhaustive search

Table 7.2 also details two commonly utilized metrics, i.e., the fraction of Pareto points (which are found in 5% to 50% of the exhaustive search time) and the coverage ratio for the total number of non-dominated design points and the significance of these points, known as the *hypervolume* indicator [ZBT07]. Although a heuristic algorithm may not always find all globally optimal solutions, the results show that our heuristic has been able to find more than 80% of the Pareto points within the first 5% of the exhaustive DSE time (as shown in Table 7.2). This Table also asserts the efficiency of X-DINC methodology in saving exploration time, which would be highly desired for the larger design spaces, where a brute-force search is not time-wise tractable. For example, simulating the accuracy of 6561 possible kernel configurations in C++ for BSS takes 312 hours (on a Rack Server equipped with Intel Xeon E5-2667 CPU at 3.20 GHz and 512 GB Random Access Memory (RAM)) when analyzed on 25 set of 4-second audio samples from the MIMII database. Please note that many *near*-Pareto points (the heuristic finds that) also render admissible resource-accuracy trade-offs but are not reported in this Table.

We have also compared the greedy strategy of [Tas+20, Jud+15] against ours. Such approaches only considered the ΔQoR as the deciding metric in navigating the heuristic (i.e., the precision of which kernel should be reduced at each step). In contrast, our approach considers the $\frac{\Delta \text{ADP}}{\Delta \text{QoR}}$ as the salience metric. We have compared

Table 7.2 Percentage of 35 Pareto points, found in different exploration times of the heuristic

Heuristic iteration (% of exhaustive search)	5	15	25	35	50
Coverage ratio of pareto points (%)	80	85.7	85.7	85.7	88.6
Coverage ratio of hypervolume (%)	61.8	69.8	69.8	69.8	72.5

these heuristics by analyzing various accuracy levels for the BSS QoR threshold. Figure 7.7 compares the ADP and memory requirement of the configurations generated by the heuristics (normalized to the respective values for the FP-based configuration). As can be seen in this figure, for most thresholds, our heuristic can achieve higher performance gain in terms of both ADP and memory (and also in a shorter searching time). The effectiveness of our heuristic is also even more pronounced when the accuracy threshold is higher: As can be seen in Fig. 7.7, the proportional gap between the heuristics (in terms of resource savings) is higher for the higher accuracy levels. Such an efficient heuristic can, therefore, obviate the need for an exhaustive search (the size of the design space increases exponentially w.r.t. the number of application kernels or nodes within the network) and expedite the selection of a Pareto or near-Pareto configuration at run-time and speed up decision-making actions.

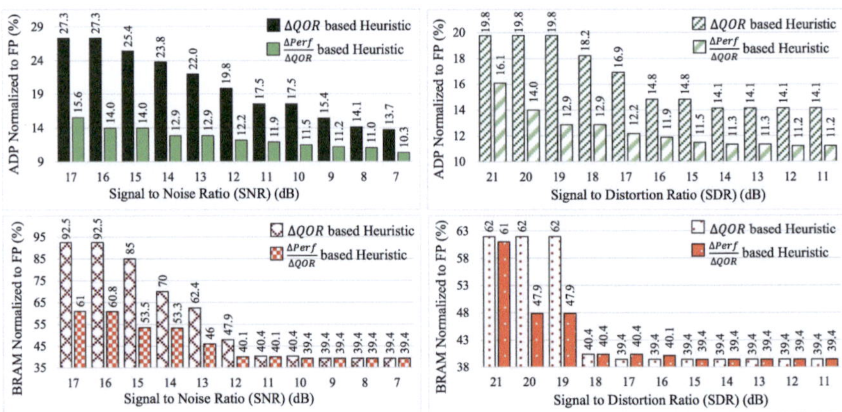

Fig. 7.7 Comparing the proposed $\frac{\Delta \text{ADP}}{\Delta \text{QoR}}$-based versus the ΔQoR-based greedy heuristic [Tas+20, Jud+15] in various QoR levels: The former (X-DINC) finds configurations with higher performance gain and lower memory requirement

Table 7.3 Grouping 8 kernels of BSS application into 2–7 nodes for distributing on network nodes in an INC scenario

Distribution scenario	Grouping	Distribution scenario	Grouping
2-Node	{1,2},{3,4,5,6,7,8}	5-Node	{1},{2},{3},{4,5},{6,7,8}
3-Node	{1},{2},{3,4,5,6,7,8}	6-Node	{1},{2},{3},{4},{5},{6,7,8}
4-Node	{1},{2},{3,4,5},{6,7,8}	7-Node	{1},{2},{3},{4},{5},{6,7},{8}

7.5.4 End-to-End Performance Gains of Distributed (INC) Over Centralized Approach

For the final evaluation of the proposed INC-tailored approximation methodology, various scenarios were considered: *Accurate-Centralized, Accurate-Distributed (INC), Approximate-Centralized,* and *Approximate-Distributed (INC).* The experimental testbed for the distributed scenarios is established by connecting one to six hops; each is equipped with a network-connected Virtex-7 FPGA. For each distribution scenario, the end-to-end latency of the application was also measured for six levels of QoR (i.e., SDR accuracy levels of 14, 16, 18, 20, 21, and FP precision as the highest accurate one). The definition of the end-to-end latency denoted as l_{e2e} of INC-accelerated scenarios for the m number of network nodes is given in Eq. 7.1. In this equation, t_c, t_p, t_u, and t_d represent the delays for computation, propagation, upload, and download, respectively[5].

$$l_{e2e} = m \cdot t_c + (m - 1) \cdot (t_u + t_p + t_d) \tag{7.1}$$

The representative values of the network's communication parameters are adopted from the 5G International Mobile Telecommunications (IMT)-2020 Key Performance Indicator (KPI) standards [Moh16] for a connected network in an industrial automation setting [Nav+20]. In this setting, the nodes are connected by links having a propagation delay of 1ms, which shows the transmission latency (t_p) between two consecutive hops. The *user experienced data rates* for downlink and uplink are also 100 and 50mbps, respectively. Note, while t_p is constant, the other variables are computed based on the size of the data that is transmitted between consecutive hops. Figure 7.8 compares the end-to-end latency of the centralized versus distributed (INC-accelerated) scenarios for different QoR-levels. Table 7.3 also shows the grouping of 8 kernels into $2 - 7$ nodes (for SDR = 20 dB). The following deductions can be inferred based on Fig. 7.8 and Table 7.3:

- First, the left part of Fig. 7.8 compares the end-to-end latency of the centralized implementations for the accurate versus approximate configurations (i.e., 14 dB $\leq$

[5] Although packing/unpacking data is also a communication overhead, evaluation of [Zhe+23a] shows it is a minor cost (<2%) in the overall traffic.

SDR $\leq$ 21 dB). The results show that the proposed approximation methodology can significantly improve the application response time, compared to the accurate approach (having 76 ms latency), while the QoR is also kept at an admissible threshold.

- Second, as shown in the right part of this figure, the associated costs of download, upload, and propagation altogether are growing linearly w.r.t. the number of network links and are much less than the computation latency. In fact, the transmission cost for 2-nodes is 2% of the total latency. It becomes at most 12% for the 7-node scenario[6]. This is attributed to the proposed approximation methodology, which 1) efficiently reduces the required number of samples and 2) reduces the bit-width of the data that needs to be transmitted between the hops by applying precision scaling. This means that the data separation job can be performed faster via the in-network acceleration through the network hops. The results also pronounce the efficacy of the proposed in-network acceleration methodology, especially when the resource footprint of one single node may not be sufficient for implementing the whole application kernels *at a high precision*: In contrast to the Accurate-Centralized FP-based implementation (which require 84 MB of memory and infeasible to be implemented in one single FPGA), all the proposed Approximate-Distributed scenarios can be practically implemented in one single FPGA, with the memory requirement of less than 30 MB per node.
- Note that the performance gains reported here, compared to the centralized approach, are based on distributing the kernels over a chain of FPGA-based nodes. The gains in the computation time become even more significant when the kernels are offloaded to other types of hardware, especially (ASIC) switches having orders of magnitude higher processing speed than a general purpose processor embedded in a stand-alone host (i.e., in a centralized approach). In fact, as discussed before, besides the computation time, the proposed approximation and distribution methodology also reduces the transmission time. Therefore, such a methodology is highly desirable, especially in distributed scenarios with more network nodes, where the transmission latency might become the dominant portion of the end-to-end latency.
- The optimal number of distributed nodes depends on application structure (e.g., kernel/operation parallelizability), required end-to-end response time, computational/memory size and node resources, and kernel output data size. X-DINC can speed up such optimization by leveraging the error-resiliency of kernels, reducing the required resources and, consequently, the number of network nodes needed to implement the target application.

[6] Communication overhead will be > 50% for the 8-node scenario, as all input samples must be transferred again to the *Tanh* kernel, if separated.

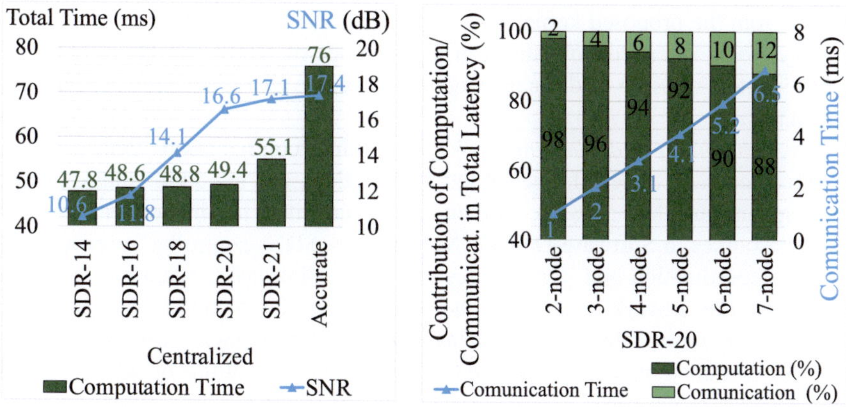

Fig. 7.8 The end-to-end latency (response time) of BSS application in the centralized (left) versus INC-accelerated and distributed approach (right)

7.6 Remarks and Challenges

7.6.1 Remarks

Scope of X-DINC: the focus of this book is narrowed to analyze the in-network acceleration of multi-kernel applications from the approximation perspective. It has been shown that by capitalizing on error-resiliency potentials across the abstraction layers within a network hierarchy, properly approximating and distributing the application kernels over the network hops can result in shorter application response time and lower resource requirements compared to the accurate centralized approach. A further analysis (e.g., on a packet-processing basis) is an interesting future track, different from the scope of this book and its claimed contributions.

Applicability to heterogeneous devices and other applications: X-DINC attempts to cope with the limitations of switches and P4 by introducing and adjusting various approximation/optimization knobs, which are agnostic from the underlying architecture. For instance, our imprecise MUL/DIV/SQRT algorithms use only shift and ADD/SUB operations. Additionally, reducing processed samples and skipping loop iterations remain hardware-agnostic. Precision scaling (4, 8, 16-bit) is supported in SoA SmartNICs from vendors like NVIDIA (BlueField), Intel (Infrastructure Processing Unit (IPU)), and Netronome. While programmable ASIC switches do not natively support this, P4 enables bit-masking and shifting to emulate lower-precision operations for both storage and computation. Similarly, loop perforation is also applicable to such devices through skipping the iterations in the so-called *recirculation* process. Thus, while the methodology was prototyped on the BSS case study and on an FPGA-based testbed, the proposed approximation techniques and methodology are applicable to other multi-kernel applications and heterogeneous network devices as well. For example, many ML algorithms, especially NNs, can significantly benefit from the proposed kernel-wise sensitivity analysis and the mixed-precision strategy (for a layer-wise quantization in the scope of NNs), considering that the multiplication, division, and square root operations are also abundantly used in such algorithms.

Task scheduling and placement: While multiple configurations with different resource-accuracy trade-offs, generated by X-DINC, can aid addressing dynamic scenarios in multi-hop networks and energy management techniques for heterogeneous resources, a Software-Defined Networking (SDN) controller is responsible for such coordination and dynamic resource allocation tasks, adapting the workloads w.r.t. the network conditions, and orchestrating distributed tasks across network infrastructures [Zhe+23b]. Such task management strategies (briefly discussed in the studies cited in Section 3.9) are not within the scope of this book.

7.6.2 Challenges

Compatibility of INC-based approach with current network setup and infrastructures: Although the preliminary results in this book and those cited in Section 3.9 have shown the advantages of INC, the question that how INC should be efficiently integrated in the current network setting (to minimize the additional overheads) is an ongoing research field for the 6G network providers. In fact, to the best of our knowledge, no existing work or simulation tool models and compares the performance of applications when implemented in an INC setup. Going into details, utilizing the existing forwarding layer to support INC is a non-trivial job[7] that is partly investigated in [KWT20, Wu+23]. Moreover, although there is currently no hardware platform that is specifically designed for in-network-based approaches, Intel has announced that its IPU carries great potential to facilitate the INC solutions [Jia+23].

Traffic management in a heterogeneous co-design approach: coupling an FPGA to an ASIC switch in a co-design approach necessitates further implementation efforts. First, the computations should be efficiently divided into two different architectures. Second, matching and merging the traffic rates (without leading to congestion) requires meticulous attention. In fact, transferring packets between an ASIC switch and its co-processor FPGA is not straightforward since the interconnect between them may act as a performance bottleneck. Such packet rates can be fixed or workload-dependent. Therefore, the switch should adapt to such dynamic scenarios. Addressing this, all switch pipelines can connect to the FPGA via a single egress port, or each pipeline can connect independently via a dedicated port [Geb+20]. Moreover, the switch also needs to differentiate between new incoming traffic and returning packets from the FPGA [Bre+20]. A few potential solutions for such concerns are already proposed in [Bre+20], which include bus adaptation and traffic encapsulation.

[7] The application computation problem is always decoupled from the data-forwarding, in literature studies.

7.7 Conclusion

This Chapter presented X-DINC, a cross-layer approximation and distribution methodology to enable or facilitate the in-network acceleration of multi-kernel applications, by capitalizing on the error-resiliency of the applications. X-DINC introduces various approximation and optimization knobs across the abstraction layers (within a network hierarchy) and efficiently adjusts them to generate multiple optimal or near-optimal configurations with acceptable performance-QoR trade-offs. Finally, X-DINC distributes the (mixed-precision) approximate kernel configurations over a set of network nodes in a way that minimizes the transmission overhead and end-to-end response time of the application (which are significantly shorter compared to the traditional accurate-centralized approach).

X-DINC is prototyped on an Xilinx Virtex-7 FPGA and evaluated on the widely used BSS application on the real-world industrial audio dataset. The evaluation results demonstrate that compared to an *Accurate-Centralized* approach, X-DINC performs the audio separation job up to 35% faster with up to 88% less ADP, respectively, when the task is distributed on 2 to 7 network nodes. These performance gains are achieved when the audio quality is remained at the acceptable range of 15-20 dB.

We believe these contributions make steps toward the in-network acceleration of multi-kernel applications, as both intermediate network nodes and IoT edge devices have heterogeneous resource footprints and processing speeds. Another potential use case for the X-DINC methodology could be an approximate-aware partitioning of the NN models, especially for an edge-cloud collaborative setting, which is a cutting-edge topic in the literature. In this context, utilizing the proposed approximation methodology can help to adjust the load of computation w.r.t the such requirements.

Chapter 8
Conclusions and Future Work

Abstract In this chapter, we first present the major conclusions of this dissertation along with some open tracks which can be further investigated as future work.

Keywords Cross-layer approximation · Reconfigurable computing · In-network computing · Distributed computing · 5G-6G · FPGA · CGRA · SIMD · MIMD · High-throughput · Pipeline · Sustainability · Energy-efficiency · Area optimization · Multi-objective heuristic · Design space exploration · Programmable networks

8.1 Book Conclusions

The past few decades have witnessed remarkable growth in the computational power of modern computing systems. This rapid increase in computational capabilities has largely been driven by the advancements in technology scaling and the development of innovative techniques and architectures deployed in the edge-to-cloud continuum. Moreover, the availability of computing platforms with low NRE cost and/or high-performance has fueled the rise and widespread adoption of computationally intensive applications. For instance, many mobile devices and edge-based embedded systems now feature dedicated processors for accelerating signal processing applications. Traditionally, the computing industry always tried to hold the strict commitment to 100% accuracy of results for all programs. However, a growing number of modern applications—ranging from various signal processing (image, video, audio, bio-signals) to ML—do not require 100% accurate computations or produce a single, definitive result. In other words, these applications exhibit an inherent resilience toward noise or approximation errors in their input data and computations. For such cases, the *approximate computing* paradigm offers a promising approach for designing high-performance and energy-efficient systems. This paradigm introduces QoR as an additional design metric, alongside resource utilization, energy consumption, latency, and throughout. In this context, Chapter 1 of this book reviews recent advancements in approximate computing, highlighting its potential and limitations.

Z. Ebrahimi and A. Kumar, *Cross-Layer Approximation and In-Network Acceleration*,
https://doi.org/10.1007/978-3-032-21712-7_8

Until recently, most of the research in approximate computing (a detailed review of which is detailed in Chapter 3) has focused on ASIC-based systems. However, the state of the arts reveals that directly applying ASIC-specific approximation techniques to FPGA-based systems often fails to deliver comparable performance gains. The primary reason lies in the architectural differences between ASICs and FPGAs . Given the increasing use of reconfigurable fabrics, especially FPGAs, to accelerate various error-resilient applications, it is imperative to develop high-throughput and energy-efficient approximation techniques tailored specifically to those architectures. Furthermore, previous approximation approaches are also posed with key drawbacks that can be classified in the following categories: Circuit-level , architecture-level, application-level, and, in general, cross-layer techniques that target both centralized and decentralized settings (for an INC environment). Below, the shortcomings of previous works in each category are summarized, followed by how this dissertation tries to address them.

8.1.1 Circuit-Level (Chap. 4)

Most prior studies have concentrated on approximating addition and multiplication, while division is shown to be the speed-limiting operation in the ALU of, e.g., CGRAs and many applications. Even works that have targeted approximation of multiplications have mainly utilized the conventional multiplication algorithms such as Baugh-Wooley, Dadda, or Booth, most of which utilize a 2D array of PPs. Sticking to an array-based design approach has, therefore, resulted in the limited performance gain in the previous approximated designs. Instead, the approximation approaches in this book utilize a logarithmic multiplication/division approach based on Mitchell's algorithms which can result in higher gains. Such a logarithmic design also significantly reduces the long latency of an accurate divider and brings it closer to that of a same-size multiplier . The proposed designs were also augmented with tunable error-reduction circuitries that efficiently bound important error metrics such as ARE and EB to $<1\%$ and 0.05%, respectively. Such metrics are cited to play a major role in minimizing the accumulated errors in multi-kernel applications having an aggregation-based structure, such as ML and signal processing (image, video, audio, and bio-signal). Finally, it is worth noting that all approximate multipliers and dividers in the literature operate exclusively in a SISD and non-pipelined mode, making them unsuitable for achieving higher throughput. Therefore, this book addresses this limitation by designing SIMD or pipelined imprecise multipliers and dividers, as discussed below.

The proposed *RAPID* designs present the first pipelined architectures for an approximate (integer) multiplier and divider. The pipelined units are designed in 2- 3-, and 4-stage variations and achieve up to $8.1\times$ higher throughput and $6.8\times$ higher throughput/Watt over their accurate IP-based counterparts from Xilinx Vivado. These variations can be adopted w.r.t the target operating frequency of FPGA and application requirement. Achieving a higher operating frequency is particularly desirable for multiplication-exhaustive applications that are fed with bulk of data.

SIMDive is the first hybrid approximate multiplier-divider in the literature, which can switch to either functionalities on-the-fly, without the need for reconfiguration. Moreover, SIMDive efficiently enables the support for data parallelism from a SIMD/MIMD perspective. Multiple configurations of SIMDive are designed that support execution at 16/8/4 or 32/16/8-bit precision. Such a SIMD/MIMD execution approach, not only enables precision-adaptability at runtime but also efficiently supports a cohesive hierarchy of approximation techniques across the layers of abstraction, from precision scaling (application-level) to inexact multiplication and division (circuit-level). Achieving a higher throughput through such a hybrid design is particularly desirable for multiplication-exhaustive applications that still encompass a few occasional division operations. That includes a broad range of parallelizable tasks that include, e.g., matrix-multiplication or convolution tasks.

The novel error-reduction schemes of SIMDive and RAPID are not only specifically tailored for reconfigurable devices, but also support various operand sizes and can be tuned to achieve different accuracy-resource trade-offs. For example, SIMDive uses a single 6-LUT to determine each bit of 64 error-reduction coefficients, and their addition to the original Mitchell design is performed with minimal overhead. Finally, in contrast to the previous works, which have mainly evaluated their designs only in a single kernel of multi-kernel applications, the efficacy of our multipliers and dividers is appraised on the end-to-end implementation of multiple multi-kernel and single-kernel applications.

8.1.2 Architecture-Level (Chap. 5)

The existing accurate and approximate CGRAs are faced with multiple architecture-level challenges. First, their PEs are built by combining separate configurations of inexact adders, multipliers, or occasionally, dividers. Such inter- or intra-PE heterogeneity has posed unavoidable overhead in logic and routing, negatively impacting the area, power consumption, and latency of the CGRAs. Second, the lack of support for division in most approximate CGRA architectures has prevented offloading division-included kernels to the CGRA and relieving the host processor from context-switching on such occasional interrupts. In fact, even in the works that have adopted an imprecise divider, the operating frequency of the ALU is very low (due to the high latency of the division, compared to the same-size multiplication). Finally, error-resilient applications exhibit varying tolerance levels to different types of precision-reduced operations (addition, multiplication, or division). As a result, a fixed precision, defined at design time, may not be suitable for all operations within a single kernel or across different application kernels. However, this challenge has been overlooked in the design of approximate CGRA, none of which support precision scaling, which can be implemented through a SIMD execution approach.

To address the aforementioned challenges, *BioCare* and *GREEN* are presented that serve as high-throughput and energy-efficient approximate CGRAs which support runtime function-versatility and precision-adaptability through the plasticity of their lightweight PEs; each can perform addition, multiplication, and division on different bit-width. By capitalizing on SIMDive multiplier-divider, the PEs in BioCare CGRA can be configured to either SISD (16-bit) or SIMD mode (sub-word of 8-bit) at runtime. The GREEN CGRA expands the function-versatility and precision-adaptability of BioCare, by enabling MIMD execution mode through supporting heterogeneous operations (ADD-MUL-DIV opcodes), on the precisions of 16-, 8-, or 4-bit. Moreover, the architecture of GREEN is equipped with a global memory, which enables measuring the reconfiguration time and energy for the end-to-end evaluation of multi-kernel applications on the GREEN CGRA. Particularly, featuring SIMD/MIMD at intra- rather than conventional inter-PE granularity mitigates the stress on global routing. Moreover, enabling each PE to execute division operations can eliminate the need for context-switching between the CGRA and the host processor when accelerating division-included kernels. BioCare enables the synergistic effects of a chain of approximation and optimization techniques within a cohesive cross-layer hierarchy: Precision scaling (application-level) on top of imprecise multiplication and division (circuit-level).

The architecture-level evaluations have demonstrated that the proposed SIMD and MIMD configurations of our CGRAs, (i.e., BioCare and GREEN, respectively) require up to 32% smaller resource footprint and 55% less energy, while enabling 49% higher processing frequency than the accurate CGRA. Moreover, the maximum throughput of the designs is also increased up to 6.7×, compared to the accurate SISD counterpart.

8.1.3 Application-Level and Cross-Layer (Chap. 6)

The existing cross-layer approximation methodologies for multi-kernel applications suffer from multiple drawbacks that have limited their end-to-end performance gains for a given accuracy threshold or, more generally, result in a poor balance between performance gains and the QoR trade-off. First, previous approaches have typically evaluated approximation on a single kernel of a multi-kernel application—sometimes without proper justification for why that specific kernel was chosen for area, power, energy, latency, or throughput efficiency. This lack of justification can be problematic, as different kernels contribute differently to each of these performance metrics. Second, the precision of operations across all kernels in the multi-kernel applications was fixed, usually at 16-bit. Consequently, the impact of precision scaling, particularly a mixed-precision approach, has not been explored in terms of the resource-accuracy trade-off. Adhering to a fixed-precision strategy in previous works ignored the fact that kernels' significance order might differ in the *gained performance* and *lost QoR* after applying the approximation.

Third, most of cross-layer strategies only focused on the domain of NNs , without exploring other application domains. In fact, the literature notably lacked a generic approximation methodology able to efficiently tune multiple approximation knobs in a multi-kernel application. Fourth, while an error-resiliency sensitivity analysis (that reveals the saliency/priority of the kernels for different approximations by measuring the trade-off between performance gain and accuracy loss in the end-to-end implementation of the application) is a pre-requisite for an efficient cross-layer approach, most of the prior research did not apply such an analysis. Some works have adopted only random error injection mechanisms, but these approaches fail to comprehensively characterize error resiliency. Finally, cross-layer studies only utilized the circuit-level techniques (e.g., approximate operators) along with application-level ones (e.g., quantization) and neglected the optimization potentials at the architecture-level (such as SIMD parallelism). Therefore, these works have been unable to effectively apply a chain of techniques within a cohesive cross-layer hierarchy. For example, in some cases, the proposed application-level optimizations could not be efficiently reflected by non-optimized hardware (e.g., applying precision scaling to multiplications in FPGA's DSPs can lead to under-utilization of these hardwired units).

This book tried to address the aforementioned challenges by proposing *Plasticine*, a cross-layer methodology for multi-kernel applications that leverages the synergistic effects of a chain of techniques across layers of abstraction (i.e., precision scaling as an application-level technique as well as the inexact configurations for multiplier and divider as circuit-level techniques). The chain of approximations in our proposed methodology is enabled in a cohesive hierarchy by utilizing the plasticity of SIMD/MIMD SIMDive multiplier-dividers. Through the use of a two-step methodology (i.e., kernel-wise error-resiliency sensitivity analysis followed by a greedy-based heuristic), Plasticine has been able to adjust the approximation knobs in multi-kernel applications in a way to maximize performance gains for a user-defined QoR.

The sensitivity analysis highlighted a key observation, which is a pre-requisite for designing an efficient cross-layer methodology in the presence of multiple techniques: Not only kernels contribute differently to each of the performance metrics (e.g., area and latency) but also their significance order might differ in the *gained performance* and *lost QoR*, after applying different approximation techniques. Our analysis recommended that for the proper adjustment of multiple knobs (i.e., adjusting the intensity of applying different approximation techniques), the *trade-off*, i.e., the relation between the end-to-end ΔPerformance-gain and ΔQoR should be considered. We have then proposed a greedy-based heuristic that properly adjusts the approximation knobs and swiftly generates configurations with the optimal- or near-optimal trade-off between performance gain and accuracy loss. Based on the observations from sensitivity analysis, the proposed heuristic adopts $\frac{\Delta\text{Performance-Gain}}{\Delta\text{QoR}}$ as the saliency/deciding metric that appropriately reflects the intensity of the gained performance over a possible accuracy loss when applying a chain of approximations. The Pareto /near-Pareto *mixed-precision* configurations generated by the heuristic enable various performance-accuracy trade-offs that render higher gains over accurate or uniform-precision counterparts in different accuracy levels.

The efficiency of the Plasticine cross-layer approximation methodology is assessed on both FPGA and CGRA and on various multi-kernel applications from different domains of application (bio-signal and image-processing to classification). The end-to-end evaluations on three multi-kernel applications employed in bio-signal processing , image processing , and moving object tracking for UAVs demonstrate 41%–64%, 39%–62%, and 70%–86% improvements in area, latency, and ADP, respectively, over 32-bit fixed precision, with negligible loss in QoR.

8.1.4 Distributed and In-Network Computing (Chap. 7)

Lastly, considering the ever-increasing complexity of applications that also require real-time responses, computations are shifting from centralized (edge or cloud) systems toward a distributed, in-network approach—i.e., processing data on the fly as it propagates through network elements. However, the heterogeneous hardware devices within the 5G networks (such as programmable switches) are resource-constrained and incapable of performing complex computations accurately and with high precision (e.g., floating-point arithmetic). Therefore, computational approaches for in-network acceleration of multi-kernel applications must be rethought and approximated, particularly while considering the limitations of programmable network devices. This cutting-edge topic has remained unexplored and is aimed to be addressed in Chap. 7.

We have presented X-DINC as a cross-layer approximation methodology for a distributed and INC setting. By further expanding and customizing the methodology of Plasticine, X-DINC identifies and effectively utilizes the synergistic effects of a chain of techniques to facilitate the deployment and distribution of kernels over heterogeneous devices in an INC environment. The identified approximation and optimization opportunities in X-DINC are identified from both the data acquisition and computation phases of multi-kernel applications. Moreover, X-DINC simplifies the computational structure of a set of complex arithmetic operations to cope with the *computation* limitations of the programmable network switches and tailor them for an INC implementation (by approximating the implementation of required yet unsupported complex arithmetic operations in programmable network switches). Finally, X-DINC distributes the (mixed-precision) approximate kernel configurations over a set of network nodes in a way that minimizes the transmission overhead and end-to-end response time of the application (which are significantly shorter compared to the traditional accurate-centralized approach).

As a proof-of-concept, X-DINC is prototyped on an industrial use case (audio separation job) for 5G/6G networks and successfully able deploy the multi-kernel application on resource-constrained network nodes. The evaluation results confirm that compared to an *Accurate-Centralized* approach, X-DINC performs the separation job up to 35% faster with up to 88% less ADP, respectively, when the task is distributed on 2 to 7 network nodes. These gains are achieved when the audio quality is maintained within the acceptable range of 15-20 dB.

8.2 Book Future Tracks

This book addresses some of the open research challenges related to the design of energy-efficient and high-throughput reconfigurable accelerators for error-resilient applications. However, there are some related research problems that provide opportunities for further research.

- *Design and exploration of higher-level approximate functional units*: More complex functional units, e.g., MAC, vector multipliers, and FP ALUs are commonly used in the PEs of various accelerators or processors such as Reduced Instruction Set Computer-V (RISC-V) soft processor. An error-configurable architecture for the aforementioned units is planned to be designed based on the proposed circuit-level techniques, aiming to improve the resource footprint and power/energy consumption of heterogeneous processing elements within the edge-to-cloud continuum. For example, our the proposed multiplier and divider designs bear a great potential to be deployed as the mantissa multiplier or divider (where the position of leading-one is fixed at MSB), as it consumes more than 95% of the total area and power in the FP unit inside the RISC-V ALU (in which division latency is up to 35× of an addition operation) [HHS18a, STa+19]. The challenge for utilizing RAPID designs relies in resolving data dependencies in consecutive instructions. Such challenges can be partially addressed through out-of-order execution in processors. Moreover, specialized versions of RAPID are targeted, which will be able to support internal data forwarding [JH19a, JH19b] and resolve data dependencies. It should be noted that bypassing would be faster and pose a smaller overhead when implemented through an intra-unit granularity.
- *Cross-architecture exploration of approximate designs*: The approximate designs presented in this book have been implemented for Xilinx FPGAs. One of the main reasons for selecting Xilinx FPGAs is their widespread utilization in both academia and industry. Our proposed designs are based on the efficient utilization of 6-input LUTs and the associated carry chains of the Xilinx FPGAs. Intel FPGAs—one of the highest market shareholders after Xilinx—also support 6-input LUTs or 8-input ALMs and associated adders to implement various types of logic. The ALM in an Intel FPGA receives 8-inputs and allows the configuration of the associated LUT in various configurations (as it was done for the 6-input LUT in Xilinx FPGAs). In fact, an 8-ALM can enable reaching a higher accuracy through the SIMDive approach: By assigning 4 MSBs of each fractional part to the inputs of one ALM, the ALM can implement one bit of the 256 error-coefficients ($2^4 \times 2^4 = 256$).
- *Utilizng the SIMD/MIMD CGRAs in an SoC framework by adding the required peripherals*: The GREEN CGRA is intended to be exploited as a co-processor in system-level frameworks such as the PULP platform [Ros+15], particularly for accelerating NN inference [Tam+21]. The SIMD/MIMD version of GREEN could serve as an interesting candidate for implementing mixed-precision quantization (e.g., 4- and 8-bit) in convolution layers. The full architectural details for the required peripherals of the SIMD/MIMD CGRAs, to be utilized in an SoC framework, will be addressed. To this end, a customized DMA needs to be designed

to allow parallel/coalesced access to/from memory and a dispatcher responsible for coordinating the simultaneous transmission of separate data at the sub-word granularity. In addition, a full-fledged CGRA mapper must be developed to map application kernels to the SIMD/MIMD PE architecture. Furthermore, measuring the communication overhead with the processor is necessary for system-level design targeting optimization of the SoC architecture. Finally, the reported performance gains for SIMDive-based architectures in this book are without considering read+pack and write+unpack overheads.[1] Indeed, a limiting factor for performance gain could be the overhead of packing/unpacking, if the data is not properly organized in the memory. It worth noting that the transmission of simultaneous and separate data at the granularity of sub-word is usually orchestrated through the controller/dispatcher. Designing a customized controller to efficiently manage SIMDization and addressing full architectural details in a system-level implementation is left for this follow-up track as well.

- *Expand the cross-layer methodology for the in-network acceleration of NN/Large Language Models (LLMs)*: The applicability of the X-DINC cross-layer approximation and distribution methodology is intended to be further expanded for accelerating the inference of NNs and LLMs in an INC environment. This cutting-edge topic is highly challenging, as a typical ASIC switch is capable to process at most a few hundreds of neurons in parallel. Therefore, executing the thousands of neurons of a single layer requires the circulation of the packet for multiple times which prolongs the latency [SSB18]. In the envisioned follow-up work, the heuristic approach of the X-DINC methodology will be optimized for utilization in a mixed-precision quantization approach. In this context, the SIMD mode of SIMDive units is identified as an interesting candidate for implementing the 4- and 8-bit convolution functions. Furthermore, a partitioning approach will be presented that properly divides a complex application into an arbitrary number of computing kernels to be offloaded to a number of network switches (determined at run time). Another interesting track could be addressing the dynamic changes in the network (e.g., when a switch or edge nodes becomes attached/detached to/from the network) with minimum reconfiguration overhead. This entails implementing a run-time coordination and management process that finds the best re-distribution scenario for the application kernels. Another interesting track could be enabling the distributed in-network training of NNs through approximation. This concept is different from the concept of federated learning from multiple perspectives, including the lack of support for computing and transmitting FP-based gradients through the ASIC-switches.
- *Run-Time adaptation of approximate accelerators via Dynamic Partial Reconfiguration (DPR)*: In some scenarios, the accuracy and performance requirements of a system may vary at run-time. For instance, a battery-powered accelerator might need to conserve energy by switching to a more energy-efficient mode, even

[1] Nevertheless, this overhead was alleviated in the proposed approach, as the kernel-wise algorithm scales operations of a kernel by the same amount and groups them in a SIMD module.

if this reduces output quality. Similarly, an accelerator could toggle between operational modes to balance throughput and output accuracy based on application demands. A potential approach to enable such run-time reconfigurability is to store various Pareto- or near-Pareto configuration of an accelerator (designed for an specific task) in the FPGA SPI/BPI flash memories [Xil18b] to be loaded at runtime for various accuracy-performance trade-offs. In this context, all parts of the multi-kernel application are not needed to be reconfigured, but a smaller portion can be reconfigured through the use of DPR technique. For example, a kernel might perform 8-bit operations to generate high-quality outputs but switch to 4-bit operations for increased energy efficiency and throughput at the cost of reduced output quality. In the 4-bit mode, the 8-bit operator could be repurposed to support multiple 4-bit operations simultaneously, thereby increasing throughput. The SIMDive design proposed in this book is a potential candidate for such run-time scenarios.

References

[Abd+19] A.A. Abdellatif, Edge computing for smart health: context-aware approaches, opportunities, and challenges. IEEE Netw 33(3), (2019)

[AC19] S. Aldossari, K..-C. Chen, Machine learning for wireless communication channel modeling: an overview. Wirel. Pers. Commun 106, (2019)

[Adv24a] Advanced Micro Devices—GitHub. Introductory examples for Vitis HLS. https://github.com/Xilinx/Vitis-HLS-Introductory-Examples/blob/master/Modeling/fixed_point_sqrt/fxp_sqrt.h (2024)

[Adv24b] Advanced Micro Devices (AMD), Introducing the T2 Card. https://www.xilinx.com/video/application/introducing-the-t2-card.html (2024)

[AGF+23] A.T.-J. Akem et al., Flowrest: practical flow-level inference in programmable switches with random forests, in *IEEE International Conference on Computer Communications (INFOCOM)* (2023)

[Akb+18] O. Akbari et al., PX-CGRA: polymorphic approximate coarse-grained reconfigurable architecture, in *Design, Automation & Test in Europe (DATE)* (2018)

[Akb+20] O. Akbari, X., CGRA: an energy-efficient approximate coarse-grained reconfigurable architecture. IEEE Trans. Comput.-Aided Des. Integr. Circ. Syst. (TCAD) (10), 39 (2020)

[All+23] D. Allan et al., *Software Defined Radio with Zynq Ultrascale+ RFSoC*, 1st edn. (Strathclyde Academic Media, 2023)

[Ama18] Amazon, Amazon Cloud Service. https://aws.amazon.com/ec2/instance-types/ (2018)

[AMD21] AMD, Divider Generator v5.1, LogiCORE IP Product Guide. https://docs.amd.com/v/u/en-US/pg151-div-gen (2021)

[AMD22] AMD-Gordon Brebner. https://opennetworking.org/wp-content/uploads/2022/05/Expanding-the-P4-universe.pdf (2022)

[Ana15] AnandTech, Samsung Exynos 7420 Deep Dive—Inside A Modern 14nm SoC. https://www.anandtech.com/show/9330/exynos-7420-deep-dive/3 (2015)

[And+21] J. Anderson et al., CGRA-ME: an open-source framework for CGRA architecture and CAD research (Invited Paper), in *IEEE International Conference on Application-specific Systems, Architectures and Processors (ASAP)* (2021)

[Anz+17] A. Anzanpour et al., Self-awareness in remote health monitoring systems using wearable electronics, in *Design, Automation & Test in Europe (DATE)* (2017)

[AR15] S. Achour, M.C. Rinard, Approximate computation with outlier detection in Topaz, in *ACM SIGPLAN International Conference on Object-Oriented Programming, Systems, Languages, and Applications* (Association for Computing Machinery, 2015)

Z. Ebrahimi and A. Kumar, *Cross-Layer Approximation and In-Network Acceleration*, https://doi.org/10.1007/978-3-032-21712-7

[Arm+22] G. Armeniakos, Hardware approximate techniques for deep neural network accelerators: a survey. ACM Comput. Surv 55(4), (2022)

[ARM21] ARM, Performance per Watt Is the New Moore's Law. https://www.arm.com/blogs/blueprint/performance-per-watt (2021)

[Asc+20] G. Ascia, Exploiting data resilience in wireless network-on-chip architectures. ACM J. Emerg. Technol. Comput. Syst 16(2), (2020). (JETC))

[Ash10] P. Ashenden, The Designer's Guide toVHDL (Morgan Kaufmann, 2010),

[Bar+15] B. Barry, Always-on vision processing unit for mobile applications. IEEE Micro 35(2), (2015)

[Bar+20] A.R. Baranwal, ReLAccS: a multi-level approach to accelerator design for reinforcement learning on FPGA-based systems. IEEE Trans. Comput.-Aided Desi. Integr. Circuit Syst. (TCAD) (2020)

[Bar20] Barefoot Networks, The world's fastest & most programmable networks. https://www.intel.com/content/www/us/en/resources-documentation/developer.html (2020)

[Bas+17] S. Basu, An inexact ultra-low power bio-signal processing architecture with lightweight error recovery. ACM Trans. Embedded Comput. Syst. (TECS) 16(5s), (2017)

[BBC15] BBC, The number glitch that can lead to catastrophe. https://www.bbc.com/future/article/20150505-the-numbers-that-lead-to-disaster (2015)

[Bec+16] A. Becher et al., A LUT-based approximate adder, in *IEEE 24th Annual International Symposium on Field-Programmable Custom Computing Machines (FCCM)* (2016)

[Beh+19] S. Behroozi et al., SAADI: a scalable accuracy approximate divider for dynamic energy-quality scaling, in *Asia & South Pacific Design Automation Conference (ASP-DAC)* (Association for Computing Machinery, 2019)

[Beh+20] R.B. Basat et al., PINT: probabilistic in-band network telemetry, in *Proceedings of the Annual Conference of the ACM Special Interest Group on Data Communication on the Applications, Technologies, Architectures, and Protocols for Computer Communication* (Association for Computing Machinery, 2020)

[BJK22] W. Byun, An energy-efficient domain-specific reconfigurable array processor with heterogeneous PEs for wearable brain-computer interface SoCs. IEEE Trans. Circuit Syst. I (TCAS-I): Reg. Pap , (2022)

[Bjö94] Å. Björck, Numerics of Gram-Schmidt orthogonalization. Linear Algebra Appl. 197–198 (1994)

[BMH14] K. Bhardwaj et al., Power- and area-efficient approximate Wallace tree multiplier for error-resilient systems, in *Fifteenth International Symposium on Quality Electronic Design* (2014)

[BMS19] A. Bosio et al., A comprehensive analysis of approximate computing techniques: from component-to application-level, in *Design, Automation & Test in Europe (DATE)* (2019)

[Bob+22] C. Bobda, The future of FPGA acceleration in datacenters and the cloud. ACM Trans. Reconfig. Technol. Syst 15(3), (2022)

[Boh09] M. Bohr, The new era of scaling in an SoC world, in *IEEE International Solid-State Circuits Conference—Digest of Technical Papers* (2009)

[Bon+18] J. Bonnot et al., A fast and fuzzy functional simulator of inexact arithmetic operators for approximate computing systems, in *ACM Great Lakes Symposium on VLSI (GLSVLSI)* (2018)

[Boo51] A.D. Booth, A signed binary multiplication technique. Q. J. Mech. Appl. Math 4(2), (1951)

[Bos+14] P. Bosshart, P4: programming protocol-independent packet processors. SIGCOMM Comput. Commun. Rev 44(3), (2014)

[Bou+07] F. Bouwens et al., *Architectural exploration of the ADRES coarse-grained reconfigurable array, in Reconfigurable Computing: Architectures, Tools and Applications* (Springer, Berlin, Heidelberg, 2007)

[Bra+17] R. Braojos, A synchronization-based hybrid-memory multi-core architecture for energy-efficient biomedical signal processing. IEEE Trans. Comput. (TC) 66(4), (2017)

[Bra+18] M. Brandalero et al., Approximate on-the-fly coarse-grained reconfigurable acceleration for general-purpose applications, in *ACM/IEEE Design Automation Conference (DAC)* (2018)

[Bre+20] P. Bressana et al., Trading latency for compute in the network, in *Proceedings of the Workshop on Network Application Integration/CoDesign* (Association for Computing Machinery, 2020)

[Bru+20] B. de Bruin et al., BrainWave: an energy-efficient EEG monitoring system—evaluation and trade-offs, in *ACM/IEEE International Symposium on Low Power Electronics and Design (ISLPED)* (Association for Computing Machinery, 2020)

[Bru+21] B. de Bruin et al., Multi-level optimization of an ultra-low power brainwave system for non-convulsive seizure detection. IEEE Trans. Biomed. Circ. Syst. (TBioCAS) **15**(5) (2021)

[BW73] C.R. Baugh, B.A. Wooley, A two's complement parallel array multiplication algorithm. IEEE Trans. Comput. C **22**, 12 (1973)

[BYB18] A. Boutros et al., Embracing diversity: enhanced DSP blocks for low-precision deep learning on FPGAs, in *IEEE International Conference on Field Programmable Logic and Applications (FPL)* (2018)

[Cas+20] J. Castro-Godínez et al., AxHLS: design space exploration and high-level synthesis of approximate accelerators using approximate functional units and analytical models, in *IEEE/ACM International Conference on Computer Aided Design (ICCAD)* (2020)

[CBD15] M. Courbariaux et al., BinaryConnect: training deep neural networks with binary weights during propagations, in *Advances in Neural Information Processing Systems*, vol. 28 (Curran Associates Inc., 2015)

[Ces+20] F.E. Rodriguez Cesen et al., Towards low latency industrial robot control in programmable data planes, in *6th IEEE Conference on Network Softwarization (NetSoft)* (2020)

[CF20] R.A. Cooke, S.A. Fahmy, Quantifying the latency benefits of near-edge and in-network FPGA acceleration, in *Proceedings of the Third ACM International Workshop on Edge Systems, Analytics and Networking* (Association for Computing Machinery, 2020)

[Cha+13] W.J. Chan et al., Statistical analysis and modeling for error composition in approximate computation circuits, in *IEEE 31st International Conference on Computer Design (ICCD)* (2013)

[Cha13] K. Chapman, Multiplexer design techniques for datapath performance with minimized routing resources (Xilinx Application Note). https://docs.xilinx.com/v/u/en-US/xapp522-mux-design-techniques (2013)

[Che+22a] K. Chen, A survey of approximate arithmetic circuits and blocks. IT Inf. Technol 64(3), (2022)

[Che+22b] X. Chen et al., Toward low-overhead inter-switch coordination in network-wide data plane program deployment, in *IEEE 42nd International Conference on Distributed Computing Systems (ICDCS)* (2022)

[Chi+13a] V.K. Chippa et al., Approximate computing: an integrated hardware approach, in *IEEE Asilomar Conference on Signals, Systems and Computers (ACSSC)* (2013)

[Chi+13b] V.K. Chippa et al., Analysis and characterization of inherent application resilience for approximate computing, in *50th Annual Design Automation Conference* (Association for Computing Machinery, 2013)

[Chi+14] V.K. Chippa, Scalable effort hardware design. IEEE Trans. Very Large Scale Integr. (VLSI) Syst 22(9), (2014)

[Chi+17] S.A. Chin et al., CGRA-ME: a unified framework for CGRA modelling and exploration, in *IEEE International Conference on Application-specific Systems, Architectures and Processors (ASAP)* (2017)

[Cho+04] K..-J. Cho, Design of low-error fixed-width modified booth multiplier. IEEE Trans. Very Large Scale Integr. (VLSI) Syst 12(5), (2004)

[Cho+11] M. Chowdhury et al., Managing data transfers in computer clusters with orchestra, in *ACM conference for Special Interest Group on Data Communications, which specializes*

in the field of communication and computer networks (SIGCOMM) (Association for Computing Machinery, 2011)
[Cis18] Cisco, Cisco Annual Internet Report—VNI Complete Forecast Highlights. https://www.cisco.com/c/dam/m/en_us/solutions/service-provider/vni-forecast-highlights/pdf/Global_Device_Growth_Traffic_Profiles.pdf (2018)
[Cis20a] Cisco, Cisco Annual Internet Report (2018–2023) White Paper. https://www.cisco.com/c/en/us/solutions/collateral/executive-perspectives/annual-internet-report/white-paper-c11-741490.html (2020)
[Cis20b] Cisco, Annual Internet Report—VNI Complete Forecast Highlights. https://www.cisco.com/c/en/us/solutions/collateral/executive-perspectives/annual-internet-report/white-paper-c11-741490.pdf (2020)
[Cis22] Cisco, Cisco Nexus SmartNIC. https://www.cisco.com/c/en/us/products/interfaces-modules/nexus-smartnic/index.html (2022)
[CJ10] P. Comon, C. Jutten, Handbook of Blind Source Separation: Independent Component Analysis and Applications (Academic, 2010),
[CMG15] H.K. Chatterjee, Real-time detection of electrocardiogram wave features using template matching and implementation in FPGA. Int. J. Biomed. Eng. Technol. (IJBET) 17(3), (2015)
[Cor14] Intel Corp., Integer Arithmetic IP Cores User Guide. https://cdrdv2-public.intel.com/705559/ug_lpm_alt_mfug-14.1-683490-705559.pdf (2014)
[Cor20a] Intel Corp., IIntel Stratix 10 Variable Precision DSP Blocks User Guide. https://www.intel.com/content/dam/www/programmable/us/en/pdfs/literature/hb/stratix-10/ug-s10-dsp.pdf (2020)
[Cor20b] Intel Corp., Intel Stratix 10 Logic Array Blocks and Adaptive Logic Modules User Guide. https://www.intel.com/content/dam/www/programmable/us/en/pdfs/literature/hb/stratix-10/ug-s10-lab.pdf (2020)
[CS05] C. Charoensak, F. Sattar, A single-chip FPGA design for real-time ICA-based blind source separation algorithm, in *IEEE International Symposium on Circuits and Systems* (2005)
[CT16] F. Cagnin, A. Torcinovich, MNIST-CNN. https://github.com/integeruser/MNIST-cnn (2016)
[Dad65] L. Dadda, Some schemes for parallel multipliers. Alta Frequenza 34, (1965)
[Dal+24] A.M. Dalloo, Approximate computing: concepts, architectures, challenges, applications, and future directions. IEEE Access 12, (2024)
[Dam+23] H.J. Damsgaard, Approximate computing in B5G and 6G wireless systems: a survey and future outlook. Comput. Netw , 233 (2023)
[Dam+24] H.J. Damsgaard, Adaptive approximate computing in edge AI and IoT applications: a review. J. Syst. Archit 150, (2024)
[DBW15] S. Das et al., Error-resilient design techniques for reliable and dependable computing. IEEE Trans. Device Mater. Reliab. **15**(1) (2015)
[Den+74] R.H. Dennard, Design of ion-implanted MOSFET's with very small physical dimensions. IEEE J. Solid-State Circuits 9(5), (1974)
[DEs+18] D. Esposito, Approximate multipliers based on new approximate compressors. IEEE Trans. Circuit Syst. I (TCAS-I): Reg. Pap 65(12), (2018)
[DGK20a] C. De la Parra et al., Improving approximate neural networks for perception tasks through specialized optimization. Future Gener. Comput. Syst. (FGCS) **113** (2020)
[DGK20b] C. De la Parra et al., ProxSim: GPU-based simulation framework for cross-layer approximate DNN optimization, in *Design, Automation & Test in Europe (DATE)* (2020)
[DHP21] R. DeVore et al., Neural network approximation. Acta Numerica **30** (2021)
[Dij22] E.W. Dijkstra, A Note on Two Problems in Connexion with Graphs (Association for Computing Machinery, 2022),
[DNT16] S. Dutt et al., A comparative survey of approximate adders, in *26th International Conference Radioelektronika (RADIOELEKTRONIKA)* (2016)

[DNT17] S. Dutt, Analysis and design of adders for approximate computing. ACM Trans. Embedded Comput. Syst. (TECS) 17(2), (2017)
[DSS20] D. Ding et al., Estimating logarithmic and exponential functions to track network traffic entropy in P4, in *IEEE/IFIP Network Operations and Management Symposium (NOMS)* (2020)
[Duc+17] L. Duch, HEAL-WEAR: an ultra-low power heterogeneous system for bio-signal analysis. IEEE Trans. Circuits Syst. I: Reg. Pap. (TCAS-I) 64(9), (2017)
[Duc+19] L. Duch, i-DPs CGRA: an interleaved-datapaths reconfigurable accelerator for embedded bio-signal processing. IEEE Embedded Syst. Lett. (ESL) **11**(2) (2019)
[Dup+20a] E. Dupuis et al., On the automatic exploration of weight sharing for deep neural network compression, in *Design, Automation & Test in Europe (DATE)* (2020)
[Dup+20b] E. Dupuis et al., Sensitivity analysis and compression opportunities in DNNs using weight sharing, in *23rd International Symposium on Design and Diagnostics of Electronic Circuits & Systems (DDECS)* (2020)
[Ebr+21] Z. Ebrahimi, Plasticine: a cross-layer approximation methodology for multi-kernel applications through minimally biased, high-throughput, and energy-efficient SIMD soft multiplier-divider. ACM Trans. Des. Autom. Electron. Syst. (TODAES) 27(2), (2021)
[Ebr+22] Z. Ebrahimi et al., RAPID: AppRoximAte pipelined soft multipliers and dividers for high-throughput and energy-efficiency. IEEE Trans. Comput.-Aided Des. Integr. Circuits Syst. (TCAD) (2022)
[ED17] A.H. El Moussawi, S. Derrien, Superword level parallelism aware word length optimization, in *Design, Automation, & Test in Europe (DATE)* (2017)
[EK21] Z. Ebrahimi, A. Kumar, BioCare: an energy-efficient CGRA for bio-signal processing at the edge, in *IEEE International Symposium on Circuits and Systems (ISCAS)* (2021)
[EK24] Z. Ebrahimi, A. Kumar, GREEN: an approximate SIMD/MIMD CGRA for energy-efficient Processing at the edge. IEEE Trans. Comput.-Aided Des. Integr. Circuits Syst. (TCAD) (2024)
[EL20] H. Englund, N. Lindskog, Secure acceleration on cloud-based FPGAs-FPGA enclaves, in *IEEE International Parallel and Distributed Processing Symposium Workshops (IPDPSW)* (2020)
[Esm+11] H. Esmaeilzadeh, Dark silicon and the end of multicore scaling. SIGARCH Comput. Archit. News 39(3), (2011)
[ESS20] E. Adams, Approximate restoring dividers using inexact cells and estimation from partial remainders. IEEE Trans. Comput. (TC) 69(4), (2020)
[EUK20] Z. Ebrahimi et al., SIMDive: approximate SIMD soft multiplier-divider for FPGAs with tunable accuracy, in *Great Lakes Symposium on VLSI* (Association for Computing Machinery, 2020)
[Fan+19a] Y. Fan et al., AxDNN: towards cross-layer design of approximate DNNs, in *Asia & South Pacific Design Automation Conference (ASP-DAC)* (ACM, 2019)
[Fan+19b] Y. Fan et al., AxDNN: towards the cross-layer design of approximate DNNs, in *24th Asia and South Pacific Design Automation Conference* (Association for Computing Machinery, 2019)
[Fan+20] H. Fan et al., VisDrone-SOT2020: the vision meets drone single object tracking challenge results, in *Workshops in European Conference on Computer Vision (ECCV)* (Springer, 2020)
[Far+20] J. Faraone, AddNet: deep neural networks using FPGA-optimized multipliers. IEEE Trans. Very Large Scale Integr. (VLSI) Syst 28(1), (2020)
[FFr+20] F. Frustaci et al., Approximate multipliers with dynamic truncation for energy reduction via graceful quality degradation. IEEE Trans. Circuits Syst. II (TCAS-II): Express Briefs **67**(12) (2020)
[FGS19] K.E. Fleming et al., Apparatus, Methods, and Systems with a Configurable Spatial Accelerator. US Patent 10,445,250 (2019)
[Fie+16] A. Fiessler et al., HyPaFilter: a versatile hybrid FPGA packet filter, in *Proceedings of the Symposium on Architectures for Networking and Communications Systems* (Association for Computing Machinery, 2016)

[Fir+18] D. Firestone et al., Azure accelerated networking: SmartNICs in the public cloud, in: *15th USENIX Symposium on Networked Systems Design and Implementation (NSDI 18)* (USENIX Association, 2018)

[Fly+95] M.J. Flynn et al., *Computer Architecture: Pipelined and Parallel Processor Design* (Jones & Bartlett Learning, 1995)

[Fra18] F. Leibiger, Computing turbo for the Open Telekom Cloud. https://www.telekom.com/de/medien/medieninformationen/detail/rechenturbo-fuer-die-open-telekom-cloud-532216 (2018)

[Fu+17a] Y. Fu et al., Deep Learning with INT8 Optimization on Xilinx Devices. White Paper (2017)

[Fu+17b] Y. Fu et al., Embedded vision with INT8 optimization on Xilinx devices. WP490 (v1.0.1), Apr 19 (2017)

[FVS15] S.A. Fahmy et al., Virtualized FPGA accelerators for efficient cloud computing, in *IEEE 7th International Conference on Cloud Computing Technology and Science (CloudCom)* (2015)

[Gai+19] B. Gaide et al., Xilinx adaptive compute acceleration platform: VersalTM architecture, in *ACM/SIGDA International Symposium on Field-Programmable Gate Arrays (FPGA)* (Association for Computing Machinery, 2019)

[Gal+20] P.P. Galuzio et al., MOBOpt—multi-objective Bayesian optimization. SoftwareX **12** (2020)

[Geb+20] N. Gebara et al., Challenging the stateless quo of programmable switches, in *19th ACM Workshop on Hot Topics in Networks. Association for Computing Machinery* (2020)

[GK17] G. A. Gillani, A.B.J. Kokkeler, Improving error resilience analysis methodology of iterative workloads for approximate computing, in *ACM Computing Frontiers Conference (CF)* (2017)

[Gle+19] R. Glebke et al., Towards executing computer vision functionality on programmable network devices, in *ACM CoNEXT Workshop on Emerging In-Network Computing Paradigms (ENCP)* (Association for Computing Machinery, 2019)

[Gob+21] L.R. Gobatto, Programmable data planes meets in-network computing: a review of the state of the art and prospective directions. J. Integr. Circuits Syst 16(2), (2021)

[Gol+00] A.L. Goldberger et al., PhysioBank, PhysioToolkit, and PhysioNet: components of a new research resource for complex physiologic signals. Circulation (2000)

[Gon+19] Y. Gong, ARA: cross-layer approximate computing framework based reconfigurable architecture for CNNs. Microelectron. J 87, (2019)

[Gon+19] X. Gu, A real-time FPGA-based accelerator for ECG analysis and diagnosis using association-rule mining. ACM Trans. Embedded Comput. Syst 15(2), (2016). (TECS))

[Gup+13] V. Gupta, Low-power digital signal processing using approximate adders. IEEE Trans. Comput.-Aided Des. Integr. Circuits Syst 32(1), (2013)

[Gup15] P.K. Gupta, Xeon+ FPGA platform for the data center, in *Fourth Workshop on the Intersections of Computer Architecture and Reconfigurable Logic (CARL)*, vol. 119 (2015)

[Han+18] M.A. Hanif, X-DNNs: systematic cross-layer approximations for energy-efficient deep neural networks. J. Low Power Electron. (JLPE) 14(4), (2018)

[Has+18] S. Hashemi et al., Approximate computing for biometric security systems: a case study on iris scanning, in *Design, Automation, & Test in Europe (DATE)* (2018)

[He+16] K. He et al., Deep residual learning for image recognition, in *IEEE Conference on Computer Vision and Pattern Recognition (CVPR)* (2016)

[He+22] J. He, Functional split of in-network deep learning for 6G: a feasibility study. IEEE Wirel. Commun 29(5), (2022)

[HHS18a] H. Saadat, Minimally biased multipliers for approximate integer and floating-point multiplication. IEEE Trans. Comput.-Aided Des. Integr. Circuits Syst. (TCAD) 37(11), (2018)

[HHS18b] M.A. Hanif et al., Error resilience analysis for systematically employing approximate computing in convolutional neural networks, in *Design, Automation, & Test in Europe (DATE)* (2018)

[HHS19] H. Saadat et al., Approximate integer and floating-point dividers with near-zero error bias, in *IEEE/ACM Design Automation Conference (DAC)* (2019)

[HJi+18] H. Jiang et al., Adaptive approximation in arithmetic circuits: a low-power unsigned divider design, in *Design, Automation & Test in Europe (DATE)* (2018)

[HJi+19] H. Jiang, Low-power unsigned divider and square root circuit designs using adaptive approximation. IEEE Trans. Comput. (TC) 68(11), (2019)

[HJi+20] H. Jiang et al., Approximate arithmetic circuits: a survey, characterization, and recent applications. Proc. IEEE **108**(12) (2020)

[HKS19] M.A. Hanif et al., CANN: curable approximations for high-performance deep neural network accelerators, in *56th ACM/IEEE Design Automation Conference (DAC)* (2019)

[HL11] J. Huang, J. Lach, Exploring the fidelity-efficiency design space using imprecise arithmetic, in *Asia & South Pacific Design Automation Conference (ASP-DAC)* (2011)

[HLR12] J. Huang et al., A methodology for energy-quality trade-off using imprecise hardware, in *IEEE/ACM Design Automation Conference (DAC)* (2012)

[HM07] S. Herbert, D. Marculescu, Analysis of dynamic voltage/frequency scaling in chip-multiprocessors, in *International Symposium on Low Power Electronics and Design* (Association for Computing Machinery, 2007)

[HMV20] D. Hodan et al., Semantically-oriented mutation operator in cartesian genetic programming for evolutionary circuit design, in *Genetic and Evolutionary Computation Conference* (Association for Computing Machinery, 2020)

[Hof+09] H. Hoffmann, Using code perforation to improve performance, reduce energy consumption, and respond to failures (Computer Science and Artificial Intelligence Laboratory Technical Report of MIT University, 2009),

[How+17] A.G. Howard, MobileNets: Efficient Convolutional Neural Networks for Mobile Vision Applications (2017),

[HS21] V.P. Ha, O. Sentieys, Leveraging Bayesian optimization to speed up automatic precision tuning, in *Design, Automation, & Test in Europe (DATE)* (2021)

[HS88] C. Harris, M. Stephens, A combined corner and edge detector, in *Alvey Vision Conference (AVC)*, vol. 15 (1988), p. 50

[HSa+20] H. Saadat et al., REALM: reduced-error approximate log-based integer multiplier, in *Design, Automation & Test in Europe (DATE)* (2020)

[Hu+21] N. Hu, An energy-efficient in-network computing paradigm for 6G. IEEE Trans. Green Commun. Netw. (TGCN) 5(4), (2021)

[HYS20] V.P. Ha et al., Towards generic and scalable word-length optimization, in *Design, Automation, & Test in Europe (DATE)* (2020)

[Hyv99] A. Hyvarinen, Fast and robust fixed-point algorithms for independent component analysis. IEEE Trans. Neural Netw 10(3), (1999)

[Iba+19] S. Ibanez et al., The P4-NetFPGA workflow for line-rate packet processing, in *ACM/SIGDA International Symposium on Field-Programmable Gate Arrays* (Association for Computing Machinery, 2019)

[Int18] Intel, Barefoot Networks-Barefoot Tofino. https://barefootnetworks.com/products/brief-tofino/ (2018)

[Int19] Intel, Intel FPGA PAC N3000 Acceleration. https://www.intel.com/content/www/us/en/software-kit/665779/intel-fpga-pac-n3000-acceleration-stacks-v1-1.html? (2019)

[Int20] Intel, Intel Stratix 10 GX/SX Device Overview. https://www.intel.com/content/www/us/en/programmable/documentation/sfo1410070178831.html (2020)

[Int21] Intel FPGAs 5G O-RU, Build More Cost-Effective and More Efficient 5G Radios with Intel Agilex FPGAs—White Paper. https://www.intel.com/content/dam/www/central-libraries/us/en/documents/build-5g-radios-with-agilex-fpgas-white-paper.pdf (2021)

[Int22] Intel, Intel Tofino 2 Specification. https://www.intel.de/content/www/de/de/products/sku/218648/intel-tofino-2-12-8-tbps-20-stage-4-pipelines/specifications.html (2022)

[Int23a] Intel, Intel FPGA Product Catalog (version 23.1). https://www.intel.com/content/dam/www/central-libraries/us/en/documents/product-catalog.pdf (2023)

[Int23b] Intel Corp., Unleash Your Data Center—Acceleration in the Data Center. https://www.intel.de/content/www/de/de/data-center/products/programmable/overview.html (2023)

[Intar] Intel—Data Center and Networking Security. Intel Agilex 7 FPGAs and Intel eASIC Devices Target IPUs, SmartNICs, and 5G Networks. https://www.intel.com/content/dam/www/central-libraries/us/en/documents/agilex-fpgas-target-ipus-smartnics-5g-networks-white-paper.pdf

[Jai+18] S. Jain et al., Compensated-DNN: energy efficient low-precision deep neural networks by compensating quantization errors, in *55th Annual Design Automation Conference* (Association for Computing Machinery, 2018)

[JH19a] J. Seo, D.H. Kim, Dependency-resolving intra-unit pipeline architecture for high-throughput multipliers, in *Design, Automation & Test in Europe (DATE)* (2019)

[JH19b] J. Seo, D.H. Kim, High-throughput multiplier architectures enabled by intra-unit fast forwarding, in *IEEE International Symposium on Computer Arithmetic (ARITH)* (2019)

[Ji+13] L. Ji et al., Design of a FPGA-based baseband for MIMO TD-LTE BTS, in *IEEE International Wireless Symposium (IWS)* (2013)

[Jia+17] H. Jiang, A review, classification, and comparative evaluation of approximate arithmetic circuits. J. Emerg. Technol. Comput. Syst. (TETC) 13(4), (2017)

[Jia+18] X. Jiao et al., Energy-efficient neural networks using approximate computation reuse, in *Design, Automation & Test in Europe Conference & Exhibition (DATE)* (2018)

[Jia+23] W. Jiang, Accelerating distributed cloud storage systems with in-network computing. IEEE Netw 37(4), (2023)

[Jin+17] X. Jin et al., NetCache: balancing key-value stores with fast in-network caching, in *ACM Symposium on Operating Systems Principles (SOSP)* (Association for Computing Machinery, 2017)

[Jin+20] C. Jin, Security of Cloud FPGAs: A Survey (2020), (arXiv preprint)

[JJ85] J. Pan, W.J. Tompkins, A real-time QRS detection algorithm. IEEE Trans. Biomed. Eng. (TBE) 32(3), (1985)

[JOA13] M. Jridi et al., Low complexity DCT engine for image and video compression, in *Real-Time Image and Video Processing (RTIVP)*, vol. 8656 (SPIE, 2013)

[Joh+20] W. John, The future of cloud computing highly distributed with heterogeneous hardware. Ericsson Technol. Rev 2020(5), (2020)

[JSD20] P.J. Edavoor, Approximate multiplier design using novel dual-stage 4:2 compressors. IEEE Access 8, (2020)

[JTT03] S..-J. Jou, Low-error reduced-width Booth multipliers for DSP applications. IEEE Trans. Circuits Syst. I: Fundam. Theory Appl 50(11), (2003)

[Jud+15] P. Judd, Reduced-precision Strategies for Bounded Memory in Deep Neural Nets (2015), (arXiv preprint)

[Jud+16] P. Judd et al., Proteus: exploiting numerical precision variability in deep neural networks, in *International Conference on Supercomputing (ICS)* (Association for Computing Machinery, 2016)

[Kam+22] M.M. Kamruzzaman et al., *Wireless communication channel modeling using machine learning* (Trans. Emerg. Telecommun, Technol, 2022)

[Kar+19] M. Karakoy, Architecture-aware approximate computing. Proc. ACM Meas. Anal. Comput. Syst 3(2), (2019)

[KGa+19] K. Gao et al., Sensitivity of multiview 3D point cloud reconstruction to compression quality and image feature detectability, in *IEEE Applied Imagery and Pattern Recognition Workshop (AIPR)* (2019)

[KGE11] P. Kulkarni et al., Trading accuracy for power with an underdesigned multiplier architecture, in *24th International Conference on VLSI Design* (2011)

[KGY10] K.Y. Kyaw et al., Low-power high-speed multiplier for error-tolerant application, in *IEEE International Conference of Electron Devices and Solid-State Circuits (EDSSC)* (2010)

[Kim+09] Y. Kim, Low power reconfiguration technique for coarse-grained reconfigurable architecture. IEEE Trans. Very Large Scale Integr. Syst. (TVLSI) 17(5), (2009)

[Kim+14] C. Kim, ULP-SRP: ultra low-power Samsung reconfigurable processor for biomedical applications. ACM Trans. Reconfig. Technol. Syst. (TRETS) 7(3), (2014)

[KK14] S.M. Kim, H.K. Kim, Direction-of-arrival based SNR estimation for dual-microphone speech enhancement. IEEE/ACM Trans. Audio Speech Lang. Process. **22**(12) (2014)

[KKS15] G. Keramidas et al., Clumsy value cache: an approximate memoization technique for mobile GPU fragment shaders, in *Workshop on approximate computing (WAPCO'15)* (2015)

[KL22] D. Kim, S. Lee, Approximate square root circuits with low latency and power dissipation. Electronics 11(1), (2022)

[Kos+17] P. Kostic et al., Heart Disease and Early Heart Attack Care. https://www.bnl.gov/hr/occmed/hpp/linkable_files/pdf/EarlyHeartAttackSymptoms.pdf (2017)

[KR07] I. Kuon, J. Rose, Measuring the gap between FPGAs and ASICs. IEEE Trans. Comput.-Aided Des. Integr. Circuits Syst. (TCAD) 26(2), (2007)

[Kri+12] A. Krizhevsky et al., Image-Net classification with deep convolutional neural networks, in *Advances in Neural Information Processing Systems (NIPS)* (2012)

[KT23] S. Kianpisheh, T. Taleb, A survey on in-network computing: programmable data plane and technology specific applications. IEEE Commun. Surv. Tutor 25(1), (2023)

[Kum+17] M. Kumm et al., Resource optimal design of large multipliers for FPGAs, in *IEEE 24th Symposium on Computer Arithmetic (ARITH)* (2017)

[Kun+21] I. Kunze et al., Investigating the applicability of in-network computing to industrial scenarios, in *4th IEEE International Conference on Industrial Cyber-Physical Systems (ICPS)* (2021)

[KWT20] I. Kunze et al., Transport protocol issues of in-network computing systems, in *Internet Engineering Task Force* (2020)

[Laf24] P. Lafaye de Micheaux, S. van der Walt, G. Varoquaux, B. Thirion, A. Gramfort, D.A. Engemann, Scikit-learn github Repository. https://github.com/scikit-learn/scikit-learn/blob/main/sklearn/decomposition/_fastica.py (2024)

[LAG19] Â.C. Lapolli et al., Offloading real-time DDoS attack detection to programmable data planes, in *IFIP/IEEE Symposium on Integrated Network and Service Management (IM)* (2019)

[Lan+05] M. Lanuzza et al., Low-cost fully reconfigurable data-path for-FPGA-based multimedia processor, in *IEEE International Conference on Field Programmable Logic and Applications (FPL)* (2005)

[Lao+21] C.L. Lao et al., ATP: in-network aggregation for multi-tenant learning, in *18th USENIX Symposium on Networked Systems Design and Implementation (NSDI 21)* (USENIX Association, 2021)

[Las+19] A.D. Lascorz et al., ShapeShifter: enabling fine-grain data width adaptation in deep learning, in *52nd Annual IEEE/ACM International Symposium on Microarchitecture* (Association for Computing Machinery, 2019)

[LB12] U. Lotrič, P. Bulić, Applicability of approximate multipliers in hardware neural networks. Neurocomputing **2012**, 96 (2012)

[LB18] M. Langhammer, G. Baeckler, High density and performance multiplication for FPGA, in *IEEE 25th Symposium on Computer Arithmetic (ARITH)* (2018)

[LCh+16] L. Chen, On the design of approximate restoring dividers for error-tolerant applications. IEEE Trans. Comput. (TC) 65(8), (2016)

[LCh+18] L. Chen, Design, evaluation and application of approximate high-radix dividers. IEEE Trans. Multi-Scale Comput. Syst. (TMSCS) 4(3), (2018)

[LDe12] L. Deng, The MNIST database of handwritten digit images for machine learning research (best of the web). ISPREG 29(6), (2012)

[Lee+10] L. Leem et al., ERSA: error resilient system architecture for probabilistic applications, in *Design, Automation & Test in Europe (DATE 2010)* (2010)

[Lee+16] S. Lee et al., Statistical quality modeling of approximate hardware, in *International Symposium on Quality Electronic Design (ISQED)* (2016)

[Leo+23] V. Leon et al., *Approximate Computing Survey* (Application-Specific & Architectural Approximation Techniques and Applications. arXiv preprint, Part II, 2023)

[LG19] S. Lee, A. Gerstlauer, Approximate high-level synthesis of custom hardware, in *Approximate Circuits: Methodologies and CAD* (Springer, 2019)

[LGB20] M. Langhammer et al., High density pipelined 8bit multiplier systolic arrays for FPGA, in *ACM/SIGDA International Symposium on Field-Programmable Gate Arrays (FPGA)* (2020)

[LHC19] A. Lerner et al., The case for network accelerated query processing, in *IEEE/ACM Conference on Innovative Data Systems Research (CIDR)* (2019)

[Li+08] Z. Li et al., FPGA design and implementation of an improved 32-bit binary logarithm converter, in *4th International Conference on Wireless Communications, Networking and Mobile Computing* (2008)

[Liu+14] C. Liu et al., A low-power, high-performance approximate multiplier with configurable partial error recovery, in *Design, Automation & Test in Europe (DATE)* (2014)

[Liu+19a] B. Liu, An ultra-low power always-on keyword spotting accelerator using quantized convolutional neural network and voltage-domain analog switching network-based approximate computing. IEEE Access 7, (2019)

[Liu+19b] L. Liu, A survey of coarse-grained reconfigurable architecture and design: taxonomy, challenges, and applications. ACM Comput. Surv. (CSUR) 52(6), (2019)

[Liu+21] Y. Liu, A survey of stochastic computing neural networks for machine learning applications. IEEE Trans. Neural Netw. Learn. Syst 32(7), (2021)

[LJ15] J.Y. L. Low, C.C. Jong, Unified Mitchell-based approximation for efficient logarithmic conversion circuit. IEEE Trans. Comput 64(6), (2015)

[LJG17] S. Lee et al., High-level synthesis of approximate hardware under joint precision and voltage scaling, in *Design, Automation, & Test in Europe (DATE)* (2017)

[LL13] C.-H. Lin, I.-C. Lin, High accuracy approximate multiplier with error correction, in *IEEE 31st International Conference on Computer Design (ICCD)* (2013)

[LL22] W. Liu, F. Lombardi, Approximate Computing (Springer, 2022),

[LLS20] W. Liu, A retrospective and prospective view of approximate computing. Proc. IEEE 108(3), (2020)

[LSF17] J. Lopes et al., Evaluation of CGRA architecture for real-time processing of biological signals on wearable devices, in *International Conference on ReConFigurable Computing and FPGAs (ReConfig)* (2017)

[LTA16] D. Lin et al., Fixed point quantization of deep convolutional networks, in *33rd International Conference on Machine Learning*, vol. 48 (PMLR, 2016)

[Lui+21] M.C. Luizelli, In-network neural networks: challenges and opportunities for innovation. IEEE Netw 35(6), (2021)

[Luo+22] J. Luo et al., Binary neural network with P4 on programmable data plane, in *IEEE International Conference on Mobility, Sensing and Networking (MSN)* (2022)

[Luo18] L.L. Luo, Towards converged SmartNIC architecture for bare metal & public clouds, in *APNet Industry Talks* (2018)

[Lyu+20] Y. Lyu, UAVid: a semantic segmentation dataset for UAV imagery. ISPRS J. Photogram. Remote Sens 165, (2020)

[Lyu+22] F. Lyu, High-throughput low-latency pipelined divider for single-precision floating-point numbers. IEEE Trans. Very Large Scale Integr. (VLSI) Syst 30(4), (2022)

[Mar14] C. Martin, Multicore processors: challenges, opportunities, emerging trends, in *Proceeing of Embedded World Conference*, vol. 14 (2014)

[Mas+24] T. Masoudi, Stochastic and approximate computing for deep learning: a survey, in *Design and Applications of Emerging Computer Systems* (Springer, 2024)

[Mel+19] J. Melchert, SAADI-EC: a quality-configurable approximate divider for energy efficiency. IEEE Trans. Very Large Scale Integr. (VLSI) Syst 27(11), (2019)

[MH21] S. Ma, J. Huai, Approximate computation for big data analytics. SIGWEB Newslett , 2021 (2021)

[Mic23] Microsoft, Unlocking the potential of in-network computing for telecommunication workloads. https://azure.microsoft.com/en-us/blog/unlocking-the-potential-of-in-network-computing-for-telecommunication-workload/ (2023)

[Mic94] G.D. Micheli, Synthesis and Optimization of Digital Circuits (McGraw-Hill Higher Education, 1994),

[Mig+16] J.S. Miguel et al., The Bunker Cache for spatio-value approximation, in *49th Annual IEEE/ACM International Symposium on Microarchitecture (MICRO)* (2016)

[Mis+14] A.K. Mishra, (2014), (iACT: T. WACAS)

[Mit16] S. Mittal, A survey of techniques for approximate computing. ACM Comput. Surv. (CSUR) **48**(4) (2016)

[MLa+19] M. Langhammer et al., Extracting INT8 multipliers from INT18 multipliers, in *IEEE International Conference on Field Programmable Logic and Applications (FPL)* (2019)

[Moc+12] Y.O.M. Moctar et al., Reducing the cost of floating-point mantissa alignment and normalization in FPGAs, in *ACM/SIGDA International Symposium on Field Programmable Gate Arrays* (Association for Computing Machinery, 2012)

[Moc74] J. Mockus, On Bayesian methods for seeking the extremum, in *International Federation for Information Processing (IFIP) Networking Conference* (Springer, 1974)

[Moh16] E. Mohyeldin, Minimum technical performance requirements for IMT-2020 radio interface(s), in *ITU-R Workshop on IMT-2020 Terrestrial Radio Interfaces* (2016)

[Mon+17] G. Montavon, Explaining nonlinear classification decisions with deep Taylor decomposition. Pattern Recognit , 65 (2017)

[Moo+17] B. Moons et al., Minimum energy quantized neural networks, in *51st Asilomar Conference on Signals, Systems, and Computers* (2017)

[Moo98] G.E. Moore, Cramming more components onto integrated circuits. Proc. IEEE 86(1), (1998)

[Mra+19] V. Mrazek et al., ALWANN: automatic layer-wise approximation of deep neural network accelerators without retraining, in *IEEE/ACM International Conference on Computer-Aided Design (ICCAD)* (2019)

[MSG16] M. Mueller et al., A benchmark and simulator for UAV tracking, in *European Conference on Computer Vision (ECCV)* (Springer, 2016)

[MSh+16] M. Shafique et al., Cross-layer approximate computing: from logic to architectures, in *53nd ACM/EDAC/IEEE Design Automation Conference (DAC)* (2016)

[MSM18] G. Montavon, Methods for interpreting and understanding deep neural networks. Digit. Signal Process 73, (2018)

[MSt+20] A.G.M. Strollo et al., Comparison and extension of approximate 4-2 compressors for low-power approximate multipliers. IEEE Trans. Circuits Syst. I (TCAS-I): Reg. Pap. **67**(9) (2020)

[MSV20] V. Mrazek, Libraries of approximate circuits: automated design and application in CNN accelerators. IEEE J. Emerg. Sel. Top. Circuits Syst. (JETCAS) 10(4), (2020)

[MSV22] M. Malekpourshahraki et al., ADA: arithmetic operations with adaptive TCAM population in programmable switches, in *IEEE 42nd International Conference on Distributed Computing Systems (ICDCS)* (2022)

[MWa+20] M. Wang et al., An optimized compression strategy for compressor-based approximate multiplier, in *IEEE International Symposium on Circuits and Systems (ISCAS)* (2020)

[Nap18] Napatech, Napatech SmartNIC Solutions Virtual Switch Acceleration. https://www.napatech.com/support/resources/solution-descriptions/napatech-smartnic-solutions-virtual-switch-acceleration/ (2018)

[Nar+15] S. Narayanamoorthy, Energy-efficient approximate multiplication for digital signal processing and classification applications. IEEE Trans. Very Large Scale Integr. Syst. (VLSI) (6), 23 (2015)

[Nav+20] J. Navarro-Ortiz, A survey on 5G usage scenarios and traffic models. IEEE Commun. Surv. Tutor 22(2), (2020)

[Nep+19] K. Nepal et al., Automated high-level generation of low-power approximate computing circuits. IEEE Trans. Emerg. Top. Comput. (TETC) **7**(1) (2019)

[Net18] Netronome, Packet and Netronome Innovate on Smart Networking-Focused Edge Compute Hardware. https://www.netronome.com/press-releases/packet-and-netronome-innovate-smart-networking-focused-edge-compute-hardware/ (2018)

[Net23a] Netronome, Agilio CX SmartNICs. https://www.netronome.com/products/agilio-cx/ (2023)

[Net23b] Netronome, Agilio CX SmartNICs. https://www.netronome.com/products/agilio-cx/ (2023)

[Nic17] C. Nicol, A Coarse Grain Reconfigurable Array for Statically Scheduled Data Flow Computing. Wave Computing White Paper (2017),

[Nie+20] G.S. Niemiec, A survey on FPGA support for the feasible execution of virtualized network functions. IEEE Commun. Surv. Tutor 22(1), (2020)

[NMi62] J.N. Mitchell, Computer multiplication and division using binary logarithms. IRE Trans. Electron. Comput. (IRETEC) 11(4), (1962)

[NMP19] E. Nogues, Algorithmic-level approximate computing applied to energy efficient HEVC decoding. IEEE Trans. Emerg. Top. Comput. (TETC) 7(1), (2019)

[Nok19] Nokia, AirScale base station product sheet, technology report. https://www.nokia.com/networks/radio-access-networks/airscale-radio (2019)

[NPH19] F. Neugebauer et al., On the limits of stochastic computing, in *IEEE International Conference on Rebooting Computing (ICRC)* (2019)

[NT17] R. Nath, D. Tullsen, Chapter 18: Accurately modeling GPGPU frequency scaling with the CRISP performance model, in *Advances in GPU Research and Practice* (2017)

[NVB16] A. Nieto, PRECISION: a reconfigurable SIMD/MIMD coprocessor for computer vision systems-on-chip. IEEE Trans. Comput. (TC) 65(8), (2016)

[NVI24a] NVIDIA, NVIDIA BlueField Data Processing Units. https://www.nvidia.com/en-us/networking/products/data-processing-unit/ (2024)

[NVI24b] NVIDIA, NVIDIA Spectrum SN4000 Open Ethernet Switches. https://www.nvidia.com/en-us/networking/ethernet-switching/spectrum-sn4000/ (2024)

[NWB22] N. Nauman et al., Real-time digital filtering for IoT data in programmable network switches, in *52nd Annual IEEE/IFIP International Conference on Dependable Systems and Networks—Supplemental Volume (DSN-S)* (2022)

[Nyq28] H. Nyquist, Certain topics in telegraph transmission theory. Trans. Am. Inst. Electr. Eng 47(2), (1928)

[OAk+17] O. Akbari, Dual-quality 4:2 compressors for utilizing in dynamic accuracy configurable multipliers. IEEE Trans. Very Large Scale Integr. Syst. (TVLSI) 25(4), (2017)

[OAk+18] O. Akbari, Toward approximate computing for coarse-grained reconfigurable architectures. IEEE Micro 38(6), (2018)

[Odo12] C. Odom, Independent component analysis algorithm FPGA design to perform real-time blind source separation (2012),

[Ope11] OpenCores, Hardware Division Unit. https://opencores.org/projects/divider (2011)

[Ope22] Open Networking Foundation (NF). https://p4.org/specs/ (2022)

[OR19] R.R. Osorio, G. Rodríguez, Truncated SIMD multiplier architecture for approximate computing in low-power programmable processors. IEEE Access 7, (2019)

[Par10] B. Parhami, *Computer Arithmetic*, vol. 20.00 (Oxford University Press, New York, NY, 2010)

[Pat+22] S. Patel et al., In-network fractional calculations using P4 for scientific computing workloads, in *Proceedings of the 5th International Workshop on P4 in Europe* (Association for Computing Machinery, 2022)

[PCL11] J.K. Paek, Binary acceleration using coarse-grained reconfigurable architecture. SIGARCH Comput. Archit. News 38(4), (2011)

[Per+04] S. Perri, Variable precision arithmetic circuits for-FPGA-based multimedia processors. IEEE Trans. Very Large Scale Integr. Syst. (TVLSI) 12(9), (2004)

[Per+06] S. Perri, SIMD multipliers for accelerating embedded processors in-FPGAs. J. Circuits Syst. Comput. (JCSC) 15(04), (2006)

[PI11] H. Parandeh-Afshar, P. Ienne, Measuring and reducing the performance gap between embedded and soft multipliers on FPGAs, in *21st International Conference on Field Programmable Logic and Applications* (2011)

[Pra+18] B.S. Prabakaran et al., DeMAS: an efficient design methodology for building approximate adders for FPGA-based systems, in *Design, Automation & Test in Europe (DATE)* (2018)

[Pra+20] R. Prasad et al., TRANSPIRE: an energy-efficient TRANSprecision floating-point Programmable archItectuRE, in *Design, Automation & Test in Europe (DATE)* (2020)

[PRS19a] B.S. Prabakaran et al., XBioSiP: a methodology for approximate bio-signal processing at the edge, in *56th ACM/IEEE Design Automation Conference (DAC)* (2019)

[PRS19b] B.S. Prabakaran et al., XBioSiP: a methodology for approximate bio-signal processing at the edge, in *ACM/IEEE Design Automation Conference (DAC)* (Association for Computing Machinery, 2019)

[PUL22] PULP-Team, FastICA algorithm for PULP platform—GitHub. https://github.com/nihil21/fast-ica-pulp/tree/main (2022)

[Pur+08] S. Purohit et al., Power-efficient high throughput reconfigurable datapath design for portable multimedia devices, in *IEEE International Conference on Reconfigurable Computing and FPGAs (ReConFig)* (2008)

[Pur+19] H. Purohit, MIMII Dataset: Sound Dataset for Malfunctioning Industrial Machine Investigation and Inspection (2019),

[Put+14] A. Putnam et al., A reconfigurable fabric for accelerating large-scale datacenter services, in *Proceeding of the Annual International Symposium on Computer Architecture* (IEEE Press, 2014)

[Qad+17] J. Qadir, Approximate networking for universal internet access. Future Internet 9(4), (2017)

[Qu+15] Y.R. Qu et al., Optimizing many-field packet classification on FPGA, multi-core general purpose processor, and GPU, in *ACM/IEEE Symposium on Architectures for Networking and Communications Systems (ANCS)* (2015)

[Que+23] H..-H. Que, A survey of approximate computing: from arithmetic units design to high-level applications. J. Comput. Sci. Technol 38(2), (2023)

[Quo+17] D.L. Quoc et al., StreamApprox: approximate computing for stream analytics, in *Proceedings of the 18th ACM/IFIP/USENIX Middleware Conference* (Association for Computing Machinery, 2017)

[Rah+15] A. Raha et al., Quality configurable reduce-and-rank for energy efficient approximate computing, in *Design, Automation & Test in Europe (DATE)* (2015)

[Rak+22] M. Rakka, Mixed-Precision Neural Networks: A Survey (2022), (arXiv preprint)

[Rak+24] M. Rakka et al., *A review of state-of-the-art mixed-precision neural network frameworks* (IEEE Trans. Pattern Anal. Mach, Intell, 2024)

[Ran+17] A. Ranjan et al., STAxCache: an approximate, energy efficient STT-MRAM cache, in *Design, Automation & Test in Europe Conference & Exhibition (DATE)* (2017)

[RBD13] M.A. Razzaque, Compression in wireless sensor networks: a survey and comparative evaluation. ACM Trans. Sens. Netw 10(1), (2013)

[RBG13] A. Rahimi, Spatial memoization: concurrent instruction reuse to correct timing errors in SIMD architectures. IEEE Trans. Circuits Syst. II Express Briefs 60(12), (2013)

[RLB20] G. Rodrigues, Survey on approximate computing and its intrinsic fault tolerance. Electronics 9(4), (2020)

[Ros+15] D. Rossi et al., PULP: a parallel ultra low power platform for next generation IoT applications, in *IEEE Hot Chips Symposium (HCS)* (2015)

[Roy+14] P. Roy et al. ASAC: automatic sensitivity analysis for approximate computing, in *ACM Conference on Languages, Compilers and Tools for Embedded Systems (LCTEC)*, vol. 49(5) (2014)

[Rüt+18] J. Rüth et al., Towards in-network industrial feedback control, in *Morning Workshop on In-Network Computing (NetCompute)* (Association for Computing Machinery, 2018)

[RWW15] P. Roy et al., PAC: program analysis for approximation-aware compilation, in *International Conference on Compilers, Architecture and Synthesis for Embedded Systems (CASES)* (2015)

[RZe+16] R. Zendegani et al., SEERAD: a high speed yet energy-efficient rounding-based approximate divider, in *Design, Automation & Test in Europe (DATE)* (2016)

[RZe+17] R. Zendegani, RoBA multiplier: a rounding-based approximate multiplier for high-speed yet energy-efficient digital signal processing. IEEE Trans. Very Large Scale Integr. Syst. (TVLSI) 25(2), (2017)

[Sam+11] A. Sampson et al., EnerJ: approximate data types for safe and general low-power computation, in *32nd ACM SIGPLAN Conference on Programming Language Design and Implementation* (Association for Computing Machinery, 2011)

[SAn+18] M.S. Ansari, Low-power approximate multipliers using encoded partial products and approximate compressors. IEEE J. Emerg. Sel. Top. Circuits Syst. (JETCAS) 8(3), (2018)

[SAn+20] M.S. Ansari, Improving the accuracy and hardware efficiency of neural networks using approximate multipliers. IEEE Trans. Very Large Scale Integr. Syst. (TVLSI) 28(2), (2020)

[San19] Sandvine, (2019), (The Global Internet-Phenomena Report)

[San21] Sandvine, *The mobile internet phenomena report* (2021), p.2021

[Sap+17] A. Sapio et al., In-network computation is a dumb idea whose time has come, in *16th ACM Workshop on Hot Topics in Networks* (Association for Computing Machinery, 2017)

[Sap+21] A. Sapio et al., Scaling distributed machine learning with in-network aggregation, in *18th USENIX Symposium on Networked Systems Design and Implementation (NSDI 21)* (USENIX Association, 2021)

[Saq+21] M. Saquetti, Toward in-network intelligence: running distributed artificial neural networks in the data plane. IEEE Commun. Lett 25(11), (2021)

[Say+15] S. Sayedsalehi, Restoring and non-restoring array divider designs in quantum-dot cellular automata. Inf. Sci , 311 (2015)

[SB18] G. Siracusano, R. Bifulco, -network Neural Networks (2018), (arXiv preprint)

[Sca+21] S. Scarfone et al., Design and analysis of a leading one detector-based approximate multiplier on FPGA, in *International Conference on Synthesis, Modeling, Analysis and Simulation Methods, and Applications to Circuit Design (SMACD)* (2021)

[Sch+17] J. Schlachter, Design and applications of approximate circuits by gate-level pruning. IEEE Trans. Very Large Scale Integr. Syst. (VLSI) 25(5), (2017)

[Sek21] L. Sekanina, *Evolutionary Algorithms in Approximate Computing: A Survey (2021)* (arXiv:2108.07000)

[Sen+19] D. Sengupta, An analytical approach for error PMF characterization in approximate circuits. IEEE Trans. Comput.-Aided Des. Integr. Circuits Syst. (TCAD) 38(1), (2019)

[SES19] S. Venkatachalam et al., Design of approximate restoring dividers, in *IEEE International Symposium on Circuits and Systems (ISCAS)* (2019)

[SF93] E.M. Schwarz, M.J. Flynn, Hardware starting approximation for the square root operation, in *IEEE International Symposium on Computer Arithmetic (ARITH)* (1993)

[SFJ19] M.S. Ansari et al., A hardware-efficient logarithmic multiplier with improved accuracy, in *Design, Automation & Test in Europe (DATE)* (2019)

[SG17] M. Shafique, S. Garg, Computing in the dark silicon era: current trends and research challenges. IEEE Des. Test 34(2), (2017)

[SGA18] C. Schorn et al., Accurate neuron resilience prediction for a flexible reliability management in neural network accelerators, in *Design, Automation, & Test in Europe (DATE)* (2018)

[SGA19] C. Schorn et al., An efficient bit-flip resilience optimization method for deep neural networks, in *Design, Automation & Test in Europe Conference & Exhibition (DATE)* (2019)

[Sha+15] M. Shafique et al., A low latency generic accuracy configurable adder, in *52nd Annual Design Automation Conference* (Association for Computing Machinery, 2015)

[Sha+16] M. Shafique et al., Invited—Cross-layer approximate computing: from logic to architectures, in *IEEE/ACM Design Automation Conference (DAC)* (Association for Computing Machinery, 2016)

[Sha+17] N.K. Sharma et al., Evaluating the power of flexible packet processing for network resource allocation, in *USENIX Symposium on Networked Systems Design and Implementation (NSDI)* (USENIX Association, 2017)

[Sha+18] S. Sharify et al., Loom: exploiting weight and activation precisions to accelerate convolutional neural networks, in *55th Annual Design Automation Conference* (Association for Computing Machinery, 2018)

[Sid+11] S. Sidiroglou-Douskos et al., Managing performance vs. accuracy trade-offs with loop perforation, in *19th ACM SIGSOFT Symposium and the 13th European Conference on Foundations of Software Engineering* (Association for Computing Machinery, 2011)

[Sir+22] G. Siracusano et al., Re-architecting traffic analysis with neural network interface cards, in *USENIX Symposium on Networked Systems Design and Implementation (NSDI 22)* (USENIX Association, 2022)

[SIS15] S.Hashemi et al., DRUM: a dynamic range unbiased multiplier for approximate applications, in *IEEE/ACM International Conference on Computer-Aided Design (ICCAD)* (2015)

[SIS16] S. Hashemi et al., A low-power dynamic divider for approximate applications, in *IEEE/ACM Design Automation Conference (DAC)* (2016)

[SKi+19] M.S. Kim, Efficient Mitchell's approximate log multipliers for convolutional neural networks. IEEE Trans. Comput. (TC) 68(5), (2019)

[SLe+19] S. Lee, Double MAC on a DSP: boosting the performance of convolutional neural networks on FPGAs. IEEE Trans. Comput.-Aided Des. Integr. Circuits Syst. (TCAD) 38(5), (2019)

[Sma18] SmartCardia, ARM Cortex-M3 inside SmartCardia (2018)

[SMS18] J. Sánchez, An analysis and implementation of the Harris corner detector. Image Process. On Line (IPOL) 8, (2018)

[Sor+07] J. Sorber et al., Eon: a language and runtime system for perpetual systems, in *5th International Conference on Embedded Networked Sensor Systems* (Association for Computing Machinery, 2007)

[SRe+16] S. Rehman et al., Architectural-space exploration of approximate multipliers, in *IEEE/ACM International Conference on Computer-Aided Design (ICCAD)* (2016)

[SS17] S. Venkatachalam, S. Ko, Design of power and area efficient approximate multipliers. IEEE Trans. Very Large Scale Integr. Syst. (TVLSI) 25(5), (2017)

[SSa+18] S. Sarwar, Energy efficient neural computing: a study of cross-layer approximations. IEEE J. Emerg. Sel. Top. Circuits Syst. (JETCAS) 8(4), (2018)

[SSA18] S. Ullah et al., SMApproxLib: library of FPGA-based approximate multipliers, in *IEEE/ACM Design Automation Conference (DAC)* (2018)

[SSB18] D. Sanvito, Can the network be the AI accelerator?, in in Morning Workshop on In-Network Computing (NetCompute). (Association for Computing Machinery, 2018),

[STa+19] S. Tamimi, An efficient SRAM-based reconfigurable architecture for embedded processors. IEEE Trans. Comput.-Aided Des. Integr. Circuits Syst. (TCAD) 38(3), (2019)

[Sta+20] P. Stanley-Marbell, Exploiting errors for efficiency: a survey from circuits to applications. ACM Comput. Surv 53(3), (2020)

[SUl+20] S. Ullah, Area-optimized accurate and approximate softcore signed multiplier architectures. IEEE Trans. Comput. (TC) , (2020)

[Sul+21] N. Sultana et al., Flightplan: dataplane disaggregation and placement for P4 programs, in *18th USENIX Symposium on Networked Systems Design and Implementation (NSDI 21)* (USENIX Association, 2021)

[SVa+17] S. Vahdat et al., TruncApp: a truncation-based approximate divider for energy efficient DSP applications, in *Design, Automation & Test in Europe (DATE)* (2017)

[SVa+19] S. Vahdat, TOSAM: an energy-efficient truncation- and rounding-based scalable approximate multiplier. IEEE Trans. Very Large Scale Integr. Syst. (TVLSI) 27(5), (2019)

[SVK07] M..-A. Song, Adaptive low-error fixed-width Booth multipliers. IEICE Trans. Fundam. Electron. Commun. Comput. Sci 90(6), (2007)

[SVM19] L. Sekanina, Automated Search-Based Functional Approximation for Digital Circuits (Springer, 2019),

[Swa+22] T. Swamy et al., Taurus: a data plane architecture for per-packet ML, in *ACM International Conference on Architectural Support for Programming Languages and Operating Systems (ASPLOS)* (Association for Computing Machinery, 2022)

[Sze+17] V. Sze et al., Efficient processing of deep neural networks: a tutorial and survey. Proc. IEEE **105**(12) (2017)

[Tam+21] S.S. P. Tammireddy, RETRACTED ARTICLE: an energy-efficient reconfigurable accelerators in multi-core systems using. PULP-NN. Appl. Nanoscience 13(3), (2021)

[Tan+19] J. Tang et al., Processors, Methods, and Systems with a Configurable Spatial Accelerator Having a Sequencer Data-flow Operator. US Patent 10,380,063 (2019)

[Tan+20] C. Tan et al., OpenCGRA: an open-source unified framework for modeling, testing, and evaluating CGRAs, in *IEEE International Conference on Computer Design (ICCD)* (2020)

[Tan+22] C. Tan et al., ASAP: automatic synthesis of area-efficient and precision-aware CGRAs, in *ACM International Conference on Supercomputing (ICS)* (Association for Computing Machinery, 2022)

[Tas+20] Z..-G. Tasoulas, Weight-oriented approximation for energy-efficient neural network inference accelerators. IEEE Trans. Circuits Syst. I: Reg. Pap 67(12), (2020)

[TMZ18] Y. Tokusashi et al., LaKe: the power of in-network computing, in *International Conference on ReConFigurable Computing and FPGAs (ReConFig)* (2018)

[TNo+20] T. Nomani, xUAVs: towards efficient approximate computing for UAVs-low power approximate adders with single LUT delay for FPGA-based aerial imaging optimization. IEEE Access 8, (2020)

[Tok+19] Y. Tokusashi et al., The case for in-network computing on demand, in *Proceedings of the Fourteenth EuroSys Conference* (Association for Computing Machinery, 2019)

[Ull+18] S. Ullah et al., Area-optimized low-latency approximate multipliers for FPGA-based hardware accelerators, in *IEEE/ACM Design Automation Conference (DAC)* (Association for Computing Machinery, 2018)

[Ull+21] S. Ullah, *High-performance accurate and approximate multipliers for FPGA-based hardware accelerators* (IEEE Trans. Comput.-Aided Des. Integr, Circuits Syst, 2021)

[Ull+22] S. Ullah, AppAxO: designing application-specific approximate operators for FPGA-based embedded systems. ACM Trans. Embedded Comput. Syst. (TECS) 21(3), (2022)

[Uni19] (University of Southern California. SIPI Image Database, 2019),

[US 25] US National Science Foundation, Cyber-Physical Systems (CPS). https://www.nsf.gov/funding/opportunities/cps-cyber-physical-systems (2025)

[USK21] S. Ullah et al., CLAppED: a design framework for implementing cross-layer approximation in FPGA-based embedded systems, in *58th ACM/IEEE Design Automation Conference (DAC)* (2021)

[VA19] E. Vahapoglu, M. Altun, From Stochastic to Bit Stream Computing: Accurate Implementation of Arithmetic Circuits and Applications in Neural Networks. arXiv preprint (2019),

[Vae+18] M. Vaeztourshizi et al., An energy-efficient, yet highly-accurate, approximate non-iterative divider, in *International Symposium on Low Power Electronics and Design (ISLPED)* (Association for Computing Machinery, 2018)

[Ven+13] S. Venkataramani et al., Quality programmable vector processors for approximate computing, in *46th Annual IEEE/ACM International Symposium on Microarchitecture* (Association for Computing Machinery, 2013)

[Ven+15a] S. Venkataramani et al., Approximate computing and the quest for computing efficiency, in *52nd Annual Design Automation Conference* (Association for Computing Machinery, 2015)

[Ven+15b] S. Venkataramani et al., Computing approximately, and efficiently, in *Design, Automation & Test in Europe (DATE)* (2015)

[Ven+16] R. Venkatagiri et al., Approxilyzer: towards a systematic framework for instruction-level approximate computing and its application to hardware resiliency, in *IEEE/ACM International Symposium on Microarchitecture (MICRO)* (2016)

[Ven+20] S. Venkataramani, Efficient AI system design with cross-layer approximate computing. Proc. IEEE 108(12), (2020)

[Ven+21a] S. Venkataramani et al., RaPiD: AI accelerator for ultra-low precision training and inference, in *IEEE International Symposium on Circuits and Systems (ISCAS)* (2021)

[Ven+21b] S. Venkataramani et al., RaPiD: AI accelerator for ultra-low precision training and inference, in *ACM/IEEE Annual International Symposium on Computer Architecture (ISCA)* (2021)

[VJ19] N.V. Toan, J. Lee, Energy-area-efficient approximate multipliers for error-tolerant applications on FPGAs, in *IEEE International System-on-Chip Conference (SOCC)* (2019)

[VL20] N. Van Toan, J. Lee, FPGA-based multi-level approximate multipliers for high-performance error-resilient applications. IEEE Access 8, (2020)

[VM17] Z. Vasicek, V. Mrazek, Trading between quality and non-functional properties of median filter in embedded systems. Genet. Program. Evolv. Mach. (GPEM) 18(1), (2017)

[VMB17] Z. Vasicek et al., Towards low power approximate DCT architecture for HEVC standard, in *Design, Automation, & Test in Europe (DATE)* (2017)

[VMr+17] V. Mrazek et al., EvoApprox8b: library of approximate adders and multipliers for circuit design and benchmarking of approximation methods, in *Design, Automation & Test in Europe (DATE)* (2017)

[Vog+19] S. Vogel et al., Self-supervised quantization of pre-trained neural networks for multiplierless acceleration, in *Design, Automation & Test in Europe (DATE)* (2019)

[Wal64] C.S. Wallace, A suggestion for a fast multiplier. IEEE Trans. Electron. Comput. EC 13(1), (1964)

[Wan+17] H. Wang et al., P4FPGA: a rapid prototyping framework for P4, in *Symposium on SDN Research* (Association for Computing Machinery, 2017)

[Wan+18] N. Wang et al., Training deep neural networks with 8-bit floating point numbers, in *Advances in Neural Information Processing Systems*, vol. 31 (Curran Associates Inc., 2018)

[Wan+22a] B. Wang et al., Towards energy-efficient CGRAs via stochastic computing, in *Design, Automation & Test in Europe (DATE)* (2022)

[Wan+22b] X. Wang, When FPGA meets cloud: a first look at performance. IEEE Trans. Cloud Comput 10(2), (2022)

[WCK21] M. Wijtvliet, CGRA-EAM-rapid energy and area estimation for coarse-grained reconfigurable architectures. ACM Trans. Reconfig. Technol. Syst. (TRETS) 14(4), (2021)

[Wij+19] M. Wijtvliet et al., Blocks: redesigning coarse grained reconfigurable architectures for energy efficiency, in *International Conference on Field Programmable Logic and Applications (FPL)* (2019)

[Wij+22] D. Wijerathne et al., Morpher: an open-source integrated compilation and simulation framework for CGRA, in *Fifth Workshop on Open-Source EDA Technology (WOSET)* (2022)

[Wik17] Wikipedia, Approximation of Euclidean distance (2017),

[Wik24] Wikipedia, https://en.wikipedia.org/wiki/IEEE_754 (2024)

[Win+20] P. Wintermeyer et al., P2GO: P4 profile-guided optimizations, in *19th ACM Workshop on Hot Topics in Networks* (Association for Computing Machinery, 2020)

[Wit+19] L. Witschen, *CIRCA: towards a modular and extensible framework for approximate circuit generation. Microelectronics Reliab* (2019), p.99

[Wit+20] R. Wittig et al., Modem design in the era of 5G and beyond: the need for a formal approach, in *27th International Conference on Telecommunications (ICT)* (2020)

[WKL11] J..-P. Wang, High-accuracy fixed-width modified booth multipliers for lossy applications. IEEE Trans. Very Large Scale Integr. (VLSI) Syst 19(1), (2011)

[WLi+18] W. Liu, Design and evaluation of approximate logarithmic multipliers for low power error-tolerant applications. IEEE Trans. Circuits Syst. I (TCAS-I): Reg. Pap 65(9), (2018)

[WLi+20] W. Liu, Design of unsigned approximate hybrid dividers based on restoring array and logarithmic dividers. IEEE J. Emerg. Sel. Top. Circuits Syst , (2020). (JETCAS)

[Wor18] World Health Organisation, Cardiovascular diseases (CVDs). https://www.who.int/news-room/fact-sheets/detail/cardiovascular-diseases-(cvds) (2018)

[WSX19] C. Wu et al., Dynamic adaptation of approximate bit-width for CNNs based on quantitative error resilience, in *IEEE/ACM International Symposium on Nanoscale Architectures (NANOARCH)* (2019)

[Wu+20a] H. Wu et al., Adaptive extraction-based independent component analysis for time-sensitive Applications, in *IEEE Global Communications Conference* (2020)

[Wu+20b] H. Wu et al., Y-Net: a dual path model for high accuracy blind source separation, in *IEEE Globecom Workshops* (2020)

[Wu+21a] H. Wu et al., Abstraction-based multi-object acoustic anomaly detection for low-complexity big data analysis, in *IEEE International Conference on Communications Workshops (ICC Workshops)* (2021)

[Wu+21b] H. Wu, Computing meets network: COIN-aware offloading for data-intensive blind source separation. IEEE Netw 35(5), (2021)

[Wu+21c] H. Wu et al., In-network processing acoustic data for anomaly detection in smart factory, in *IEEE Global Communications Conference (GLOBECOM)* (2021)

[Wu+21d] H. Wu et al., In-network processing for low-latency industrial anomaly detection in softwarized networks, in *IEEE Global Communications Conference (GLOBECOM)* (2021)

[Wu+22a] H. Wu et al., Demonstration of in-network audio processing for low-latency anomaly detection in smart factories, in *IEEE 19th Annual Consumer Communications & Networking Conference (CCNC)* (2022)

[Wu+22b] H. Wu et al., Experimental proof of the energy advantage of in-network intelligence, in *International Conference on Electrical, Computer, Communications and Mechatronics Engineering (ICECCME)* (2022)

[Wu+22c] H. Wu et al., You only hear once: lightweight in-network AI design for multi-object anomaly detection, in *IEEE 21st Mediterranean Electrotechnical Conference (MELECON)* (2022)

[Wu+23] H. Wu, Accelerating industrial IoT acoustic data separation with in-network computing. IEEE Internet Things J 10(5), (2023)

[Wu+24] Y. Wu, A survey on approximate multiplier designs for energy efficiency: from algorithms to circuits. ACM Trans. Des. Autom. Electron. Syst. (TODAES) 29(1), (2024)

[WZW21] Q. Wang, A survey on policies, modelling and security of cyber-physical systems in smart grids. Energy Convers. Econ 2(4), (2021)

[Xav+21] B.M. Xavier et al., Programmable switches for in-networking classification, in *IEEE Conference on Computer Communications (INFOCOM)* (2021)

[Xav+22] B.M. Xavier, MAP4: a pragmatic framework for in-network machine learning traffic classification. IEEE Trans. Netw. Serv. Manag 19(4), (2022)

[Xia+17] H. Xiao, Fashion-MNIST: A Novel Image Dataset for Benchmarking Machine Learning Algorithms (2017), (arXiv preprint)

[Xil13] Xilinx, Xilinx 7 SeriesFPGA Programmable Guide forHDL Designs. https://www.xilinx.com/support/documentation/sw_manuals/xilinx14_7/7series_hdl.pdf (2013)

[Xil15] Xilinx, LogiCOREIP multiplier v12.0. https://www.xilinx.com/support/documentation/ip_documentation/mult_gen/v12_0/pg108-mult-gen.pdf (2015)

[Xil16a] Xilinx, 7 Series FPGAs Configurable Logic Block—User Guide. https://docs.amd.com/v/u/en-US/ug474_7Series_CLB (2016)

[Xil16b] Xilinx, LogiCOREIP Divider v5.1. https://www.xilinx.com/support/documentation/ip_documentation/div_gen/v5_1/pg151-div-gen.pdf (2016)

[Xil17a] Xilinx, UltraScale Architecture Configurable Logic Block. https://docs.amd.com/v/u/en-US/ug574-ultrascale-clb (2017)

[Xil17b] Xilinx Corp, Xilinx Powers Huawei FPGA Accelerated Cloud Server. https://www.prnewswire.com/news-releases/xilinx-powers-huawei-fpga-accelerated-cloud-server-300514320.html (2017)

[Xil18a] Xilinx, 7 Series DSP48E1 Slice. https://www.xilinx.com/support/documentation/user_guides/ug479_7Series_DSP48E1.pdf (2018)

[Xil18b] Xilinx, 7 Series FPGAs Configuration User Guide (UG470). https://www.xilinx.com/support/documentation/user_guides/ug470_7Series_Config.pdf (2018)

[Xil18c] Xilinx, Zynq-7000 SoC Technical Reference Manual. UG585. https://www.xilinx.com/support/documentation/user_guides/ug585-Zynq-7000-TRM.pdf (2018)

[Xil20] Xilinx Wired and Wireless Group, Xilinx 5G Telco Accelerator Cards. https://www.xilinx.com/publications/presentations/xilinx-5g-telco-accelerator-cards.pdf (2020)

[Xil21] Xilinx, Vivado Design Suite User Guide: Power Analysis and Optimization. https://www.xilinx.com/content/dam/xilinx/support/documents/sw_manuals/xilinx2021_2/ug907-vivado-power-analysis-optimization.pdf (2021)

[Xil23] Xilinx, Alveo Network Accelerators. https://www.xilinx.com/applications/data-center/network-acceleration.html (2023)

[Xil24] Xilinx, Vitis High-Level Synthesis User Guide (UG1399). https://docs.amd.com/r/en-US/ug1399-vitis-hls/HLS-Math-Library?tocId=evTCuuTJjRDKk3~TAmqNmw (2024)

[XMK16] Q. Xu, Approximate computing: a survey. IEEE Des. Test 33(1), (2016)

[Xu+18] R. Xu et al., VideoChef: efficient approximation for streaming video processing pipelines, in *USENIX Annual Technical Conference (USENIX ATC 18)* (USENIX Association, 2018)

[Xu+23] W. Xu et al., ClickINC: in-network computing as a service in heterogeneous programmable data-center networks, in *Proceedings of the ACM SIGCOMM 2023 Conference* (Association for Computing Machinery, 2023)

[XZ19] Z. Xiong, N. Zilberman, Do switches dream of machine learning? Toward in-network classification, in *ACM Workshop on Hot Topics in Networks (HotNets)* (Association for Computing Machinery, 2019)

[Yan+13] Z. Yang et al., Approximate XOR/XNOR-based adders for inexact computing, in *13th IEEE International Conference on Nanotechnology (IEEE-NANO)* (2013)

[Yaz+16] A. Yazdanbakhsh, RFVP: rollback-free value prediction with safe-to-approximate loads. ACM Trans. Archit. Code Optim. (TACO) 12(4), (2016)

[Yaz+17] A. Yazdanbakhsh, AxBench: a multiplatform benchmark suite for approximate computing. IEEE Des. Test 34(2), (2017)

[Yaz+18] A. Yazdanbakhsh et al., GANAX: a unified MIMD-SIMD acceleration for generative adversarial networks, in *ACM/IEEE Annual International Symposium on Computer Architecture (ISCA)* (2018)

[YHS20] Y. Guo et al., Small-area and low-power FPGA-based multipliers using approximate elementary modules, in *Asia & South Pacific Design Automation Conference (ASP-DAC)* (2020)

[Yin+17] S. Yin et al., Minimizing area and energy of deep learning hardware design using collective low precision and structured compression, in *51st Asilomar Conference on Signals, Systems, and Computers* (2017)

[Yin+18] P. Yin et al., Design of dynamic range approximate logarithmic multipliers, in *ACM Great Lakes Symposium on VLSI (GLSVLSI)* (2018)

[YJL20] L. Yu et al., Efficient FFT implementation on a CGRA, in *International Conference on Mathematics and Artificial Intelligence (ICMAI)* (Association for Computing Machinery, 2020)

[Yoo+13] J.W. Yoon, Architecture customization of on-chip reconfigurable accelerators. ACM Trans. Des. Autom. Electron. Syst. (TODAES) 18(4), (2013)

[YYa+19] Y. Yang et al., FPNet: customized convolutional neural network for FPGA platforms, in *IEEE International Conference on Field-Programmable Technology (ICFPT)* (2019)

[ZBT07] E. Zitzler et al., *The hypervolume indicator revisited: on the design of pareto-compliant indicators via weighted integration, in Evolutionary Multi-Criterion Optimization* (Springer, Berlin, Heidelberg, 2007)

[Zer+21] G. Zervakis et al., Approximate computing for ML: state-of-the-art, challenges and visions, in *Asia & South Pacific Design Automation Conference (ASP-DAC)* (Association for Computing Machinery, 2021)

[Zhe+22a] C. Zheng, Automating In-network Machine Learning (2022), (arXiv preprint)

[Zhe+22b] C. Zheng, IIsy: Practical In-network Classification (2022), (arXiv preprint)

[Zhe+23a] C. Zheng et al., DINC: toward distributed in-network computing, in *Proc. ACM Netw. 1.CoNEXT3* (2023)

[Zhe+23b] C. Zheng et al., *In-network machine learning using programmable network devices: a survey* (IEEE Commun. Surv, Tutor, 2023)

[Zho+18] S. Zhou, DoReFa-Net: Training Low Bitwidth Convolutional Neural Networks with Low Bitwidth Gradients (2018), (arXiv preprint)

[Zho+20] Y. Zhou et al., Flow event telemetry on programmable data plane, in *Annual Conference of the ACM Special Interest Group on Data Communication on the Applications, Technologies, Architectures, and Protocols for Computer Communication* (Association for Computing Machinery, 2020)

[Zil+14] N. Zilberman, NetFPGA SUME: toward 100 Gbps as research commodity. IEEE Micro 34(5), (2014)

[ZSA20] Z. Ebrahimi et al., LeAp: leading-one detection-based softcore approximate multipliers with tunable accuracy, in *Asia & South Pacific Design Automation Conference (ASP-DAC)* (2020)

[ZZ13] Y. Zhang, Y. Zhao, Modulation domain blind speech separation in noisy environments. Speech Commun 55(10), (2013)

Index

Z. Ebrahimi and A. Kumar, *Cross-Layer Approximation and In-Network Acceleration*,
https://doi.org/10.1007/978-3-032-21712-7

MIX
Papier aus verantwortungsvollen Quellen
Paper from responsible sources
FSC® C105338

If you have any concerns about our products,
you can contact us on
ProductSafety@springernature.com

In case Publisher is established outside the EU,
the EU authorized representative is:
Springer Nature Customer Service Center GmbH
Europaplatz 3, 69115 Heidelberg, Germany

Printed by Libri Plureos GmbH
in Hamburg, Germany